First View inside an Atom— Encounters with Gerhard Richter between Art and Science

Franz J. Gießibl

Chair for Quantum-Nanoscience
Institute for Experimental and Applied Physics
University of Regensburg,
Universitätsstraße 31, D 93053 Regensburg, Germany

Verlag der Buchhandlung
Walther und Franz König, Köln
2022

Copyediting: Anja Huber
Graphic design: Silke Fahnert, Uwe Koch, Köln
Production: Druckerei Kettler, Bönen

First published by Verlag der Buchhandlung Walther und Franz König
Ehrenstraße 4, D-50672 Köln

Supported by Universitätsstiftung Regensburg

REGENSBURGER UNIVERSITÄTS STIFTUNG

Bibliographic information published by the Deutsche Nationalbibliothek
The Deutsche Nationalbibliothek lists this publication
in the Deutsche Nationalbibliografie; detailed bibliographic data
are available in the Internet at http://dnb.d-nb.de.

Printed in Germany

Vertrieb / Distribution:

Europe
Buchhandlung Walther König
Ehrenstraße 4, D-50672 Köln
Tel: +49 (0) 221 / 20 59 6 53
verlag@buchhandlung-walther-koenig.de

UK & Ireland
Cornerhouse Publications Ltd. – HOME
2 Tony Wilson Place, UK – Manchester M15 4FN
Tel: +44 (0) 161 212 3466
publications@cornerhouse.org

Outside Europe
D.A.P. / Distributed Art Publishers, Inc.
75 Broad Street, Suite 630, USA – New York, NY 10004
Tel: +1 (0) 212 627 1999
orders@dapinc.com

ISBN 978-3-7533-0188-4

Table of contents

I. Preface

Art and science share a common source: human creativity. Creating something new virtually out of nothing, as described by physics Nobel laureate Gerd Binnig in his book on creativity [1], is also perceived by some people as a threat. After all, the new often replaces something that already exists, and it is not uncommon for pioneers to be held in low esteem by those they drive to change and evolve. So what is good about creativity and why should we think about connections between art and science?

Energy and evolution have developed the Earth into a paradise over about five billion years, and in principle this could continue for a long time. Over the course of a year, the sun irradiates us with an amount of energy that, according to current rates, corresponds to about one thousand times the annual income of an employee of an industrialized nation for each citizen of the earth. The matter of which we and the world are made of are atoms. Atoms do not wear out and, at about fourteen billion years, are almost as old as the universe.

One goal of human creativity could be to design the cycles on our planet so that the overall system has a maximum lifespan. This goal requires a dialogue across disciplines, generations, and continents. Art ignites imagination and creativity, appeals to all the senses and reflects the world in a very special way. Science, on the other hand, explains the world and opens up avenues for action.

Half a millennium ago, it was still possible to be a pioneering artist and a scientist at the same time. Leonardo da Vinci was an artist, scientist, inventor, and engineer all in one [2]. Today, it is difficult to be a true universal genius because of the progress in all these fields. All the more important is the dialogue between the disciplines.

The eminent artist Gerhard Richter is interested not only in painting, but also in scientific imaging. A report in the *Frankfurter Allgemeine Zeitung* about the first-ever imaging of the charge clouds of an atom in 2000 led to our first contact. Richter delighted me, stimulated me, surprised me, and convinced me of the importance of art also in natural science. Over the last twenty years, many interactions and mutual inspirations resulted, which are reported in this book.

II. Introduction

A. First View

In the evening hours of December 30, 1999, an unusual image appeared on the computer screen of my newly developed atomic force microscope. The new instrument was able to image even smaller details than its predecessors because it was based on a force sensor invented shortly before, which was able to convert the precision of time measurement with quartz clocks into highly accurate force measurements. Line by line, the image of a silicon surface built up on the screen. Until that day, the sharpest images you could get of an atom with the most advanced microscopes appeared as more or less washed-out spheres. Now, suddenly, internal details could be seen in the image of a single atom—two kidney-shaped structures.

The new images were a minor sensation—for the first time, patterns within an atom could be made visible. The technical article appeared in July 2000, and in the same month the German daily newspaper *Frankfurter Allgemeine Zeitung* (FAZ) and weekly magazine *Der Spiegel* reported on the result.

The FAZ article titled *First View Into an Atom* aroused the interest of Gerhard Richter, one of today's most important visual artists, whose extensive work is described in the recently published biography *"Das Denken ist beim Malen das Malen"* by Armin Zweite [3]. Richter created a slightly enlarged excerpt of the FAZ article titled "Erster Blick [CR112] (2000)" shown in Figure 1, which led to a first contact between Richter and me. Over the years, our acquaintance has brought forth even further projects reported here.

B. Painting

I came into contact with painting at an early age. My mother was an enthusiastic hobby painter, she drew and painted still lifes, landscapes, portraits, traditional Bavarian wardrobes, marksmen's targets, maypole figures. To help feed her seven children, she even painted glass bottles with a moun-

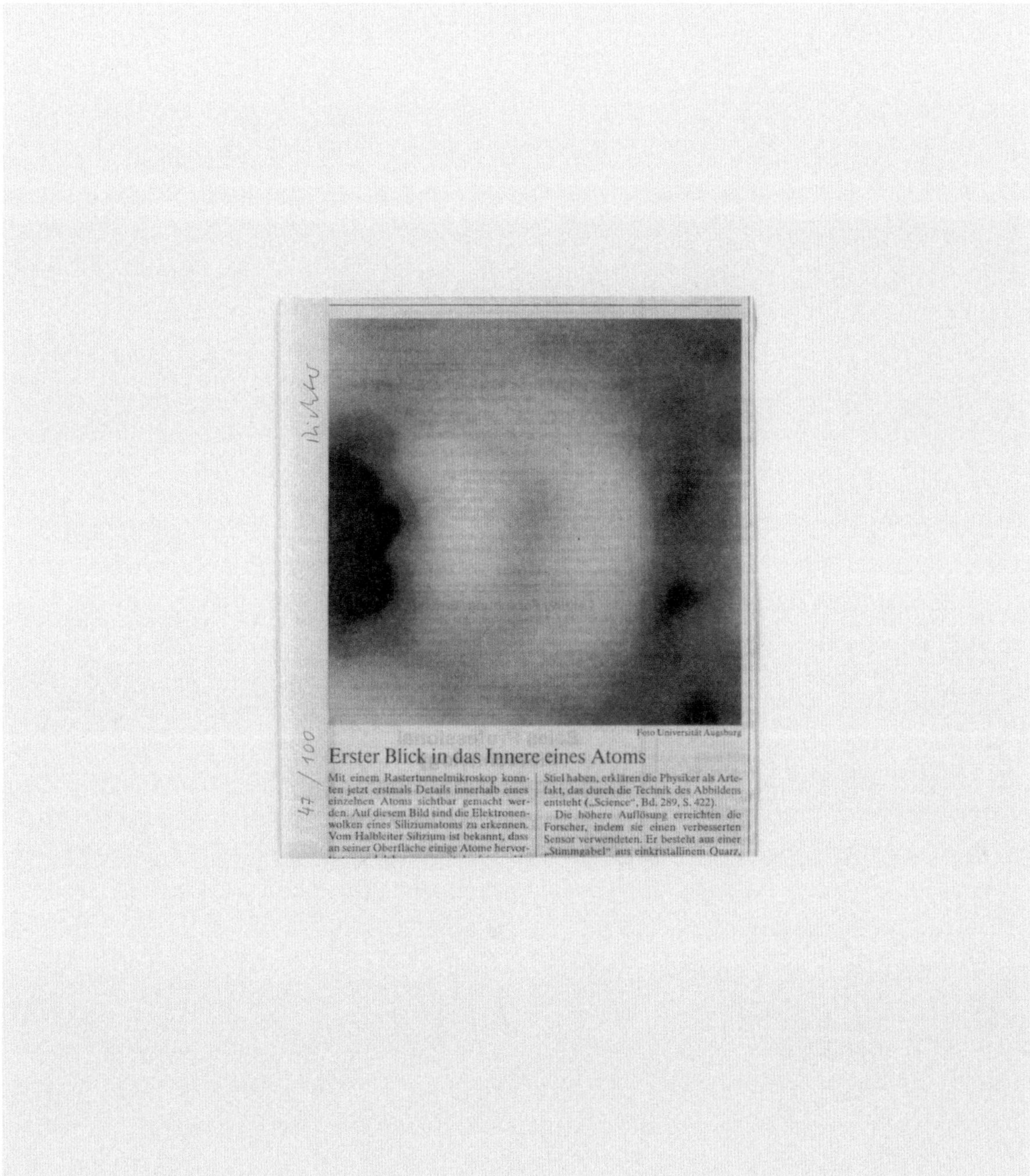
Foto Universität Augsburg

Erster Blick in das Innere eines Atoms

Mit einem Rastertunnelmikroskop konnten jetzt erstmals Details innerhalb eines einzelnen Atoms sichtbar gemacht werden. Auf diesem Bild sind die Elektronenwolken eines Siliziumatoms zu erkennen. Vom Halbleiter Silizium ist bekannt, dass an seiner Oberfläche einige Atome hervor-

Stiel haben, erklären die Physiker als Artefakt, das durch die Technik des Abbildens entsteht („Science", Bd. 289, S. 422).

Die höhere Auflösung erreichten die Forscher, indem sie einen verbesserten Sensor verwendeten. Er besteht aus einer „Stimmgabel" aus einkristallinem Quarz,

Figure 1. Gerhard Richter, First View (2000), Catalogue Raisonne CR112.

tain gentian motif for a nearby liqueur factory at home, 0.50 DM or about 0.25 EUR a piece.

Influenced by this early contact with art, I loved to visit the museums of the great masters on later trips and, if possible, their studios, such as Paul Cézanne's in Aix-en-Provence. I liked the subject of art at school, technical drawing and construction, and even took a course in figurative drawing at the Eidgenössische Technische Hochschule (Swiss Federal Institute of Technology). However, despite diligent efforts, I managed only a few halfway satisfactory drawings and watercolors. Photography seemed to be easier, so after photography courses at school and two semesters of scientific photography at the Technical University of Munich, I even set up my own photo lab with a few friends. It was clear to me that my talent lay more on the technical side than on the artistic side. However, the recognition of my own limits in painting talent was accompanied by a great respect for visual artists.

Another aspect of drawing opened up to me as a developer of microscopes —the drawing of construction plans and electronic circuit diagrams as a creative act for devices that were then manufactured, in some cases hundreds of times, and used in many parts of the world.

In a good twenty years of engagement and exchange with Richter, I discovered commonalities between experimentation and research in science and in art. Richter often remarked that the photograph is superior to the painting in many ways. For a painter, it is no longer enough to be able to paint perfectly photorealistically in order to develop painting further. Similarly, while it is gratifying for an experimental scientist to create the highest resolution microscope, attaining fulfilment requires to find an object of study that opens new doors of scientific perception.

The scientific publication on "subatomic" resolution appeared in July 2000 in the journal *Science* [4]. A short time later I received mail from a colleague in Mainz: "I have seen that your experiments have already found their way into the art scene". He sent me a catalog of Edition Klaus Staeck, where signed and numbered prints of "Erster Blick" were offered for 1200 DM in an edition of 100. I was speechless at first. Our experiment had created a little stir in the scientific world, some responses were enthusiastic, others doubted the result. In 2001 a commentary even appeared in *Science* with the remark that the whole thing could also be an artifact, a measurement error [5]. It was also criticized that the experimental data was not backed by calculations using the most modern theory, density functional theory. Although that calculation was made up three years later with most modern methods [6] and could completely confirm our simpler calculation, doubts remained with individual colleagues about the validity of our interpretation.

I was all the more pleased that Gerhard Richter used the newspaper article about our result as a template. Considering that Gerhard Richter reads the *Frankfurter Allgemeine Zeitung* and that the development of a new microscopy technique can lead with high probability to images which are also discussed in the daily press, the contact between Richter and me was therefore much more probable than, for example, the certain arrangement of the panels in Richter's painting "25 Colors" with its tiny probability of 6×10^{-26}.

After seeing Klaus Staeck's brochure, I called him by phone because I was curious why Richter had chosen that picture. Mr. Staeck said it was nice of me to call, Richter wanted to contact me anyway to find out exactly how these images were created. So Richter invited me to visit him in his studio when I was nearby. On the sidelines of a lecture tour to Bad Honnef in 2002, we arranged a first personal meeting in Richter's former studio in Bismarckstraße in Cologne, where his wife Sabine Moritz took a photo of us (Figure 2). Richter had long since become a world star by this time, and his matter-of-fact and modest demeanor, coupled with his strong interest in image-making beyond painting and photography, made a lasting impression on me.

At this first personal encounter in October 2002, Richter gave me the recently published biography *"Gerhard Richter, Maler"* by Dietmar Elger [7]. In this richly detailed work, I learned that Richter had already studied art in Dresden and that independence would have been open to him in East Germany even as a young painter, thanks to many commissions. Nevertheless, he made the profound decision to leave East Germany in an adventurous way and to move to the West. Once there, he began a new study of painting in Düsseldorf and laid the foundation for an unprecedented world career.

I found Richter's curriculum vitae very captivating, because my own also contains several apparent breaks, each of which provided great triggers for growth. It was clear to me that I could learn a great deal from him, so I was very keen to maintain contact. Richter's fascination with the scientific side of imaging the world ensured that he also kept the connection alive. From my perspective, there were certain parallels between Richter's world and my own. Richter, whose eyes still radiate a strong vividness and receptivity even at an advanced age, sees the world in a special way and opens it up to us with his thousands of works of art. Even today, he attends in detail to exhi-

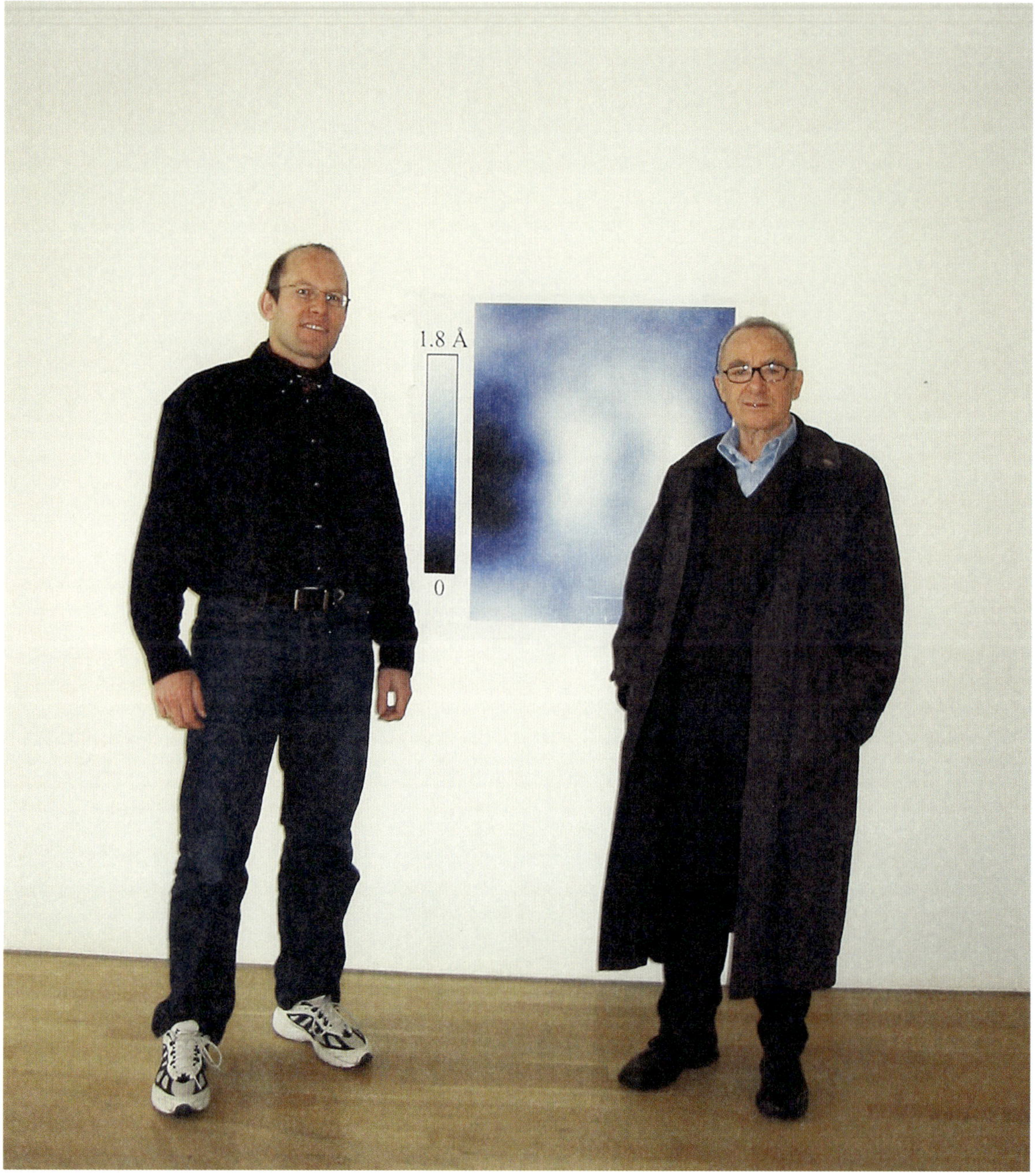

Figure 2. Gerhard Richter (right) and the author at Richter's studio in Bismarckstraße in Cologne in front of a poster of the template for "First look" with 1 billion times magnification of the silicon atom (October 22, 2002). Photo: Sabine Moritz.

bitions all over the world, planning the shows by arranging magnetically held miniature versions of the works of art on replica models of the museums and exhibition spaces in question. He has invited me to many exhibition openings, and conversely I often reported to him on conference trips and new scientific results. Richter's artworks can be seen as vehicles of visual perception, and our newly created microscopes and imaging techniques certainly have analogies.

The following chapters report on our encounters in more or less chronological order.

III. Chance

A. 25 Colors

How did this encounter come about? It was based on a cascade of coincidences. Gerhard Richter has always been fascinated by the power of chance. His painting "25 Colors", reproduced in Figure 3, shows 25 square panels of different colors, each measuring about 10 cm by 10 cm, so the whole painting measures about half a meter square. Assuming that the arrangement of the colors is random, the question arises as to the number of different ways to arrange the color panels. Intuitively, one would say that there are many, but a manageable number.

In fact, it is easy to calculate the number of possibilities. If you arrange the panels one by one, you have 25 possibilities for the first one. For the second one you have 24, for the third 23 and so on. In the end, you get 25 times 24 times 23 ... times 2 times 1 possibilities, the mathematicians have introduced as an abbreviation of this arithmetic rule the "factorial", expressed by an exclamation mark after the number, i.e. 25! This number is huge, with 25! = 15 511 210 043 330 985 984 000 000 $\approx 1.6 \times 10^{25} \approx 1/6 \times 10^{-26}$. So the probability for a given color constellation is only about 6×10^{-26}. If all variants of "25 Colors" were placed side by side, the arrangement would span a width of ten thousand times the diameter of our galaxy, the Milky Way. The probability for the certain arrangement of the colors

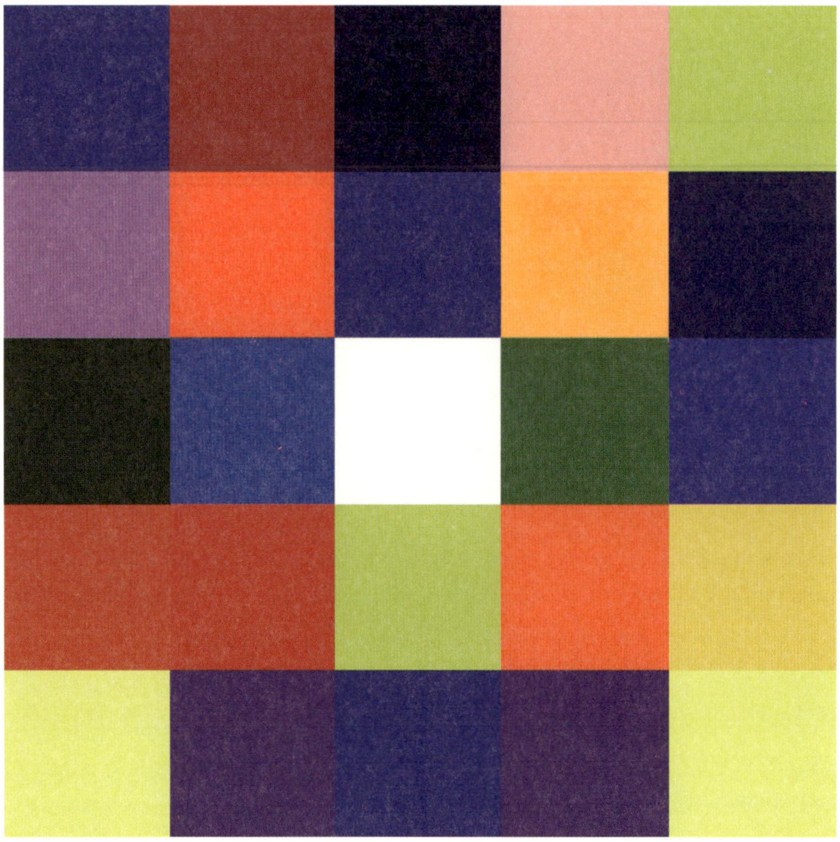

Figure 3. Gerhard Richter, 25 Colors (2007), CR [901-4].

from Richter's "25 Colors" is thus far smaller than that Gerhard Richter and I met completely by chance. If one considers the common interest in the illustration of nature and the world, the probability for a meeting increases immensely.

B. Time and what holds the world together

Technology has fascinated me for as long as I can think. As a child, I wondered why a toy that you break doesn't hold up just as well as before after the parts have been precisely reassembled. What exactly is it that actually holds matter together?

A second important question was the nature of time and its measurement with clocks. Taking a clock apart to find out its function held a great fascination for me even as a child. The composition of the finely crafted gears in Figure 4 evokes that science and art are working together here. The questions raised by Einstein about the problem of simultaneity, the nature of light, and relativity came on top of this curiosity.

If you take a clock apart and analyze the interaction of the cogs, the function of the clock becomes fully apparent. Later, I wondered whether by looking at the smallest stable components of the world, the atoms, one can also deduce the big picture. Did the reader of the FAZ article "First view inside an atom" also expect to find little gears inside and then was disappointed to see only fuzzy clouds? Richter probably felt confirmed by the result, probably that's why he created "First View".

Understanding atoms and their interactions has led to technological developments that have dramatically changed our lives. Transistors, lasers and many other developments have emerged from this and created the basis for a new world of technical possibilities. Will we one day also under-

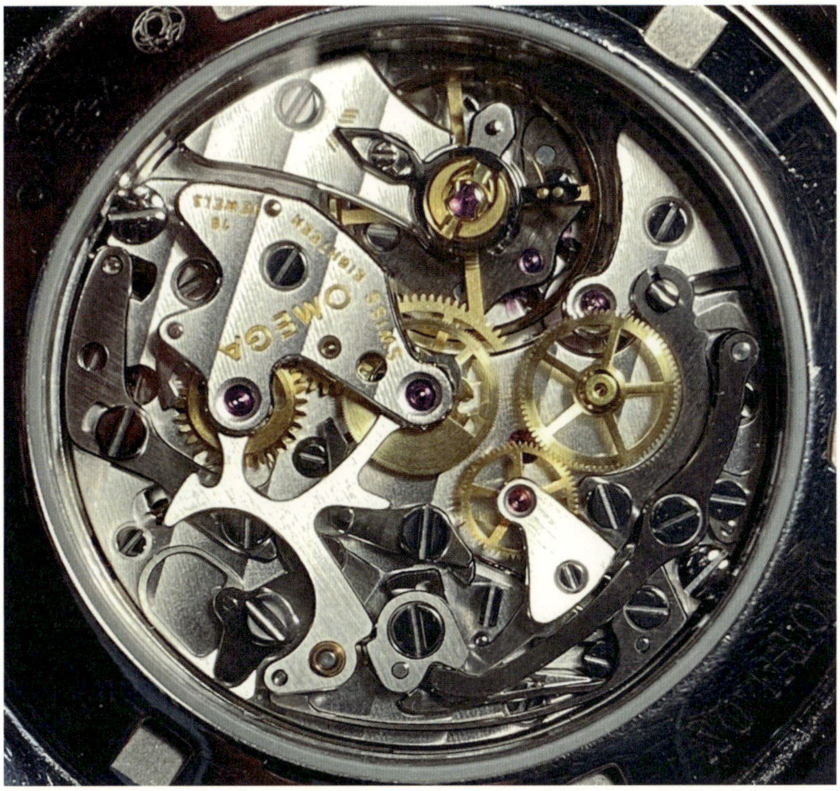

Figure 4. Movement of a mechanical hand wound wristwatch photographed through the transparent back cover.

stand the origin of living beings, including humans, on such an atomic scale?

My journey into this cosmos of questions began early, already at the age of six I irreversibly disassembled a defective mechanical alarm clock into its components with the help of two large stones. At least, the larger gears could still be used as spinning tops.

C. Chance and the path to "First View"

At the age of 15 I was allowed to accompany my father in his truck to transport a mysterious machine the size of a VW Beetle from the manufacturer located near my birthplace Amerang in Upper Bavaria to a paint shop. The machine was a tire testing machine based on holographic laser technology.

While loading the machine in the manufacturer's lab, I spotted laser light for the first time in my life and was immediately excited. When sunlight hits a wall, you see a bright spot that barely changes when you tilt your head. With coherent laser light falling on a wall, you notice a striking graininess that changes a lot when you move your head. Fascinating!

A few days later, I introduced myself to the founder of this company to ask for an internship. In the years that followed, I often spent the summer school breaks working in this holographic lab and it was clear that I had to study optics to understand all the mysteries of this fascinating laser radiation.

During my first year of studies in precision engineering and optics at the Munich University of Applied Sciences, my physics professor advised me to transfer to the Technical University because my deep thirst for knowledge could be better satisfied there. Studying physics at the TU Munich provided answers to many of my fundamental questions and powerful tools for solving hard scientific and technical problems.

In my fourth year of study, I moved to Switzerland for a year as a guest student at the Swiss Federal Institute of Technology in Zurich (ETHZ). I was captivated by the writings and life of Albert Einstein, so going to his former place of education and work was an obvious step. During the spring semester break, I again looked for an internship. I had heard in the corridors of the ETHZ that the scientists Gerd Binnig and Heinrich Rohrer had invented a new type of microscope at the IBM (International Business Machines Corporation) research laboratory in Rüschlikon, near Zurich, that can image atoms one by one. I sent them an informal application and was offered the opportunity to work on a Fabry-Pérot laser spectrometer from February to March 1986 with the task to optimize it.

My supervisor, Eric Courtens, was pleased with my work and asked me on the last day of my internship if I already had plans for the future. He told me that Gerd Binnig was currently working at Stanford in the U.S. but was planning to start a research group at Ludwig-Maximilians-Universität in Munich and was looking for coworkers. I enthusiastically agreed to his question whether I would be interested in participating. At that time, Binnig and Rohrer were already being discussed as Nobel Prize candidates and I had not dared to dream that such a possibility could come true.

A few months later, Madame Brunner, the landlady of my student room in Zurich Oerlikon, knocked on the door: "There's a phone call for you!". Gerd Binnig was on the line, and based on Eric Courten's recommendation, offered me to work with him as a student of the IBM Physics Group Munich at LMU. He recommended that I finish my diploma studies as soon as possible and offered me a PhD position for the time after that.

In the fall of 1986, I was back at the TU Munich and started a diploma thesis with Professor Gerhard Abstreiter on the growth and laser spectroscopic measurement of an artificial semiconductor made of single atomic layers of silicon and germanium. The topic was also very exciting and I learned much from Gerhard Abstreiter whow was a highly gifted teacher in experimental semiconductor physics.

In October 1986 it was announced that the 1986 Nobel Prize in Physics would be awarded half to Ernst Ruska for the scanning electron microscope he had invented half a century earlier and half to Gerd Binnig and Heinrich Rohrer for the invention of the scanning tunneling microscope [8]. The very next year, another Nobel Prize in Physics went to two researchers from IBM Rüschlikon: Georg Bednorz and Karl Alexander Müller were awarded for the discovery of ceramic high-temperature superconductors. Incidentally, Georg Bednorz had done research on strontium titanate in his diploma thesis with Müller; this material will be mentioned later.

Despite the guest year in Zurich, I was one of the first of my class of 1982 to have the diploma in my pocket and, after an extended trip through Mexico and Guatemala in the footsteps of Max Frisch's *"Homo Faber"* [9] to take up the doctoral position with Gerd Binnig in May 1988. In Frisch's novel, the initially highly technical worldview of the main character, Walter Faber, an engineering graduate of the ETH Zurich, undergoes a dramatic transformation. Faber finds himself thrown into a world where chance and forces beyond predictable science exert a powerful influence. In this respect, the trip and the study of Frisch's novel were a good preparation for what was to come.

Actually, I had not aspired to a doctorate. Hermann Lübbe, with whom I took a semester of philosophy at the University of Zurich, had remarked "universities are experience-diluted environments" and experience was exactly what I was hungry for. On the other hand, a PhD position funded by IBM was very attractive. Already the internship at IBM Rüschlikon was very professionally organized, the company gave me the feeling from the first day that my work there really counted.

Gerd Binnig proposed three doctoral themes for me to choose from. The first was an attempt to experimentally prove the gravitational waves that had been postulated by Albert Einstein in 1911. The second topic was to read out the sequence of base pairs of deoxyribonucleic acid (DNA) with a scanning tunneling microscope, i.e. to decipher the code of all life, the genetic material. The third topic proposal was the endeavor to create an atomic force microscope with which atomic resolution is possible.

The first topic was spectacularly solved in 2016 by a world-spanning team of one thousand scientists [10]. Barry Barish, Rainer Weiss and Kip Thorne were awarded the Nobel Prize in Physics and the Kavli Prize in Astrophysics for this great achievement in 2016 on behalf of the entire team.

The second topic, the decoding of DNA structure, was also solved by an elaborate and huge effort, the Human Genome Project, which also involved about a thousand people. However, this decoding has not yet been achieved with the scanning tunneling microscope, which shows that Gerd Binnig has always been a fan of pursuing very challenging goals.

The third problem, to improve the atomic force microscope such that it obtains atomic resolution, immediately appealed to me the most—the occupation with it finally answered the fundamental question from my childhood, why a broken toy does not stick again when the parts are exactly put together.

The atomic force microscope (AFM) was the new microscope Binnig had worked with in 1985 and 1986 with Stanford professor Calvin Quate and Christoph Gerber, Binnig's technician from the tunneling microscope days in Rüschlikon.

The atomic force microscope is an extension of the scanning tunneling microscope, it was invented by Binnig in 1985 [11] and the first technical realization [12] was introduced shortly thereafter together with Christoph Gerber and Calvin Quate. Its function will be explained in the next chapter.

It was due to Christoph Gerber's organizational talent and drive that the IBM Physics Group Munich at the Ludwig-Maximilians-University Munich could start quickly. Gerber received several honorary doctorates and worked as a professor at the University of Basel after leaving IBM.

The IBM physics group consisted of only a handful of employees, spatially embedded in the chair of laser expert Theodor Hänsch, who had moved from Stanford to Munich almost simultaneously with Binnig. The location of the laboratory in the heart of Schwabing, Munich's center of nightlife, was not ideal in terms of vibrations from passing buses, but it had the advantage that after a long evening in the laboratory, one could plunge directly into the action. Although Theodor Hänsch was very interested in the newly emerging atomic-resolution microscopy, he continued to focus on his field of expertise, optics. In 2005, he was awarded the Nobel Prize in Physics, along with John Hall and Roy Glauber, for pioneering precision optical measurements.

In addition to mentoring several collaborators and students in the fields of scanning probe microscopy, Binnig set a very ambitious new goal: to develop a new understanding of the evolution of nature and mankind. To this end, he lectured at the university and wrote a widely acclaimed book entitled *"Aus dem Nichts—Über die Kreativität von Natur und Mensch"* (*"Out of Nothing—On the Creativity of Nature and Man"*) [1].

My topic was comparatively simple, but still difficult enough. The problem with the atomic force microscope was that it could not initially achieve the atomic resolution I had hoped for. It turned out that the physical challenges were much more difficult than with the scanning tunneling microscope [13]. The topic of my PhD thesis was now to realize the atomic resolution of the force microscope. An even more ambitious goal would have been to resolve the complex 7 × 7-reconstruction of the silicon surface with the force microscope, because this would establish the force microscope as a powerful tool in surface physics. This goal was achieved with the scanning tunneling microscope only a few years after its invention [14] and the goal now was to first achieve atomic resolution on some surface with our force microscope built together with Gerd Binnig and Christoph Gerber in Munich. This proved to be very difficult because surfaces are chemically very reactive in a vacuum and the spring beam entered into a strong chemical bond with its scanning tip, which made non-destructive scanning impossible.

Incidentally, this experimental observation also provides the answer to the question of why a toy does not simply stick together again after it has been broken and reassembled. When the toy is broken in air, a monomolecular film of adsorbates forms on the fractured surface just a nanosecond later, preventing the parts from bonding strongly together chemically. A nanosecond is a short time, a billionth of a second, in which light travels a distance of merely 30 cm. In ultra-high vacuum, the surfaces remain clean for hours or even days; the opposite problem there is the strong adhesion of components to each other.

Because of the strong chemical bond between the tip of the force microscope and the sample, we initially chose the chemically inert surface of potassium bromide, a salt that has the same crystal structure as common rock salt [15]. The road to atomic resolution was paved with many failures and hard work, even Binnig was often present in the laboratory late at night. After a few years, the breakthrough finally came. I still remember the goose bumps and the cold shiver down my spine when the atoms of the potassium bromide surface finally appeared on the screen. It must have been a Saturday, Theodor Hänsch was experimenting in his laboratory next door and I was able to show him excitedly and enthusiastically live “the atoms” as presented in Figure 5 [16].

The next step took me to California in the summer of 1992. Stanford University was an important center for scanning probe microscopy in the 1980s and 1990s because of Binnigs sabbatical there. Sang-il Park, a native of South Korea, had developed one of the first scanning tunneling microscopes as a doctoral student with Calvin Quate at Stanford, which could reproduce the atomic image of the 7 × 7-reconstruction of the silicon surface outside the IBM laboratory in Rüschlikon. Sang-il Park subsequently founded Park Scientific Instruments in Mountain View in the heart of Silicon Valley, which he used to produce and market the tunneling microscope he had developed. He offered me a position to develop an atomic force microscope for operation in vacuum. Professor Calvin Quate, who served on the board of the company, kept visiting me in the company’s laboratory and encouraged me to try atomic imaging of the silicon surface with the new force microscope.

We were able to avoid the problem of the tip sticking to the surface, which had been observed during my doctoral studies in Munich, by setting the spring beam into resonant oscillation and using the change in oscillation frequency caused by the forces acting on the tip as the measurement signal. In spring 1994, the microscope was ready, and shortly thereafter we were the first in the world to achieve atomic resolution of the silicon surface [17]. Even the *Frankfurter Allgemeine Zeitung* reported on the result, albeit only as a text [18].

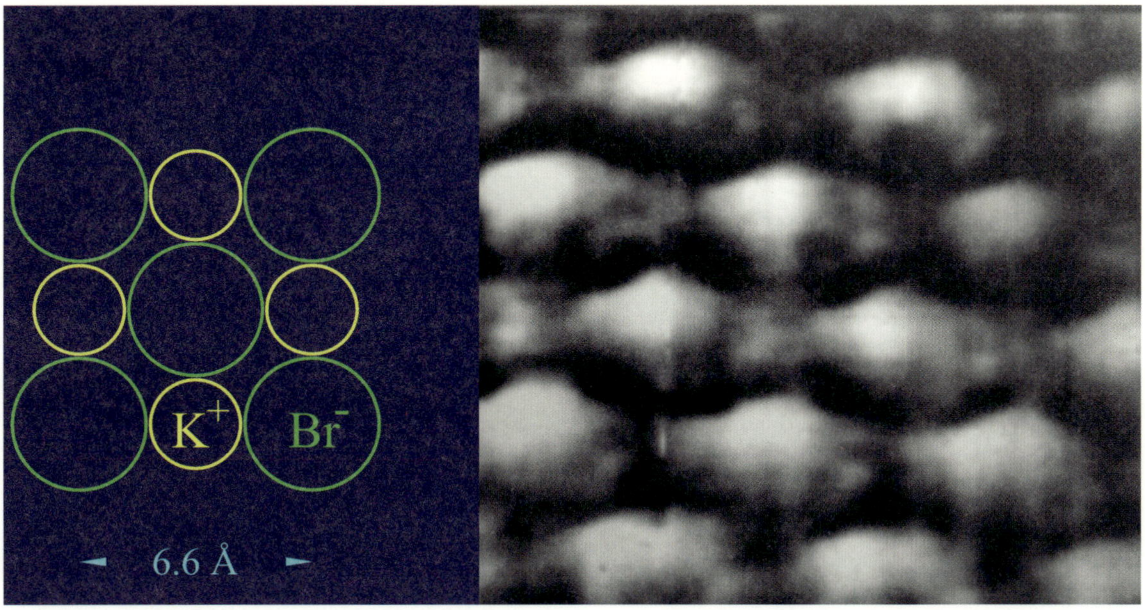

Figure 5. The (100) crystal surface of potassium bromide, atomically resolved by AFM.

With the atomic resolution of silicon, an important and difficult scientific goal had been achieved, which at times seemed unattainable even in the opinion of my very optimistic PhD advisor Gerd Binnig. The vacuum force microscope I had developed at Park Scientific Instruments in California had reached an important milestone, and I had no further scientific goal in sight. After completing the development of the microscope, the focus of my tasks moved from research and development to marketing and sales. I felt that there was a lot to learn in this field in order to successfully take innovations into global markets. Therefore, I strived to gain more experience about business and applied to the international management consulting firm McKinsey&Company in the German office, only to return to Germany after two and a half years in the USA.

This change involved a bit of a culture shock, because the dynamics and cosmopolitanism of Silicon Valley strongly contrasted with life in Germany. On the one hand, the campus of Stanford University radiated great tranquility; on the other hand, many innovations sprang from it that have changed the lives of all mankind. That is why I visited Quate again and again in this world center of high technology, even years after having returned to Germany.

I had actually finished with physics and was looking forward to new challenges in top management consulting. My expectation was that I might be able to invent new approaches and models for innovation and revenue growth programs in this new environment. The work as a consultant was very exciting, but also very demanding, because there was a lot at stake for the clients and for us consultants, and relevant results were expected after only a short time. The daily routine was correspondingly tough—getting up at five on a Monday, arriving at the client's office on the first plane, often working until midnight, and usually working from home in the McKinsey office in Munich on Friday. Because of my technical background, I was often involved in benchmarking studies where we compared critical functions of our client with the practices of global champions.

I wondered if, in my earlier work as a scientist, I had indeed always looked far enough for the world's best solution to a technical problem. In fact, chance was at play in the development of the most critical part of the force microscope, the force spring. Gerd Binnig was doing research in Professor Quate's group at Stanford during his sabbatical year in 1986. At Stanford, the microstructuring of silicon was a widely used practice, and so it was decided that the cantilevers that sensed the atomic forces were to be made of silicon. The benchmarking experience at McKinsey stimulated me to think about whether silicon was really the optimal material. After all, we had measured the sensing force by the change in frequency of the silicon cantilever.

The measurement of time and frequency is a centuries-old problem of watchmakers [19]. In fact, there are no watches that measure time by the oscillation frequency of a silicon bar, although lately some manufacturers have been using silicon in balance springs for some years. In contrast, the introduction of quartz watches in the 1970s caused a revolution in the watch industry. Even Rolex, a premium manufacturer of ornate mechanical watches, produced watches with quartz movements for a few years. In general, quartz watches proved to be extremely precise, inexpensive and low-maintenance. The timing element of the quartz watch is a quartz tuning fork, which is a few millimeters in size and whose frequency is highly stable (Figure 6). Therefore, the question arose whether it would be possible to make much more precise spring beams from quartz. When I informed the partners of McKinsey&Company about this idea, the company granted me paid and unpaid leave to investigate this question in my rapidly established home laboratory. In fact, using a quartz tuning fork as the basis for a force sensor yielded very promising results from the start, and so I filed a patent application for what I called the "qPlus Sensor" as an improvement on the silicon cantilever (see Figure 6).

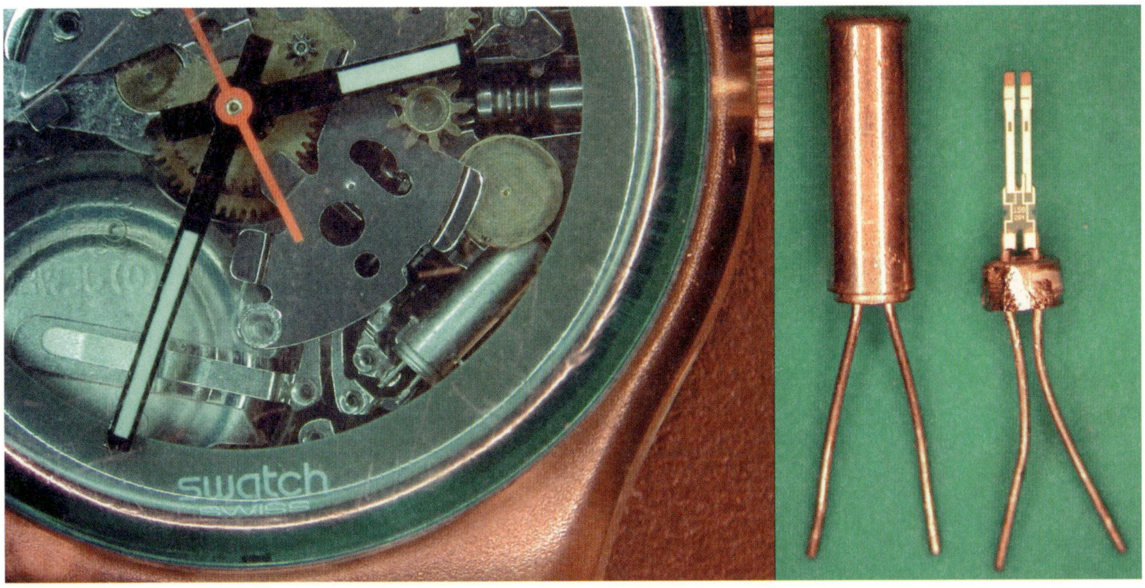

Figure 6. Left: Photo of a transparent Swatch watch from 1986. The slanted metal cylinder near the 4 o'clock position contains the clock generator of the watch. Right: The encapsulated quartz generator consists of an evacuated metal cylinder containing a quartz tuning fork. This quartz tuning fork oscillates at a frequency of exactly 32 768 Hz, i.e. 2^{15} Hz, generating an AC current that can be sensed via the two wires. An electronic circuit in the clock divides the clock signal 15 times by 2 and then arrives at the cycle of one second, which drives the clock's hand. Due to the high frequency stability and the directly generated electrical signal, the quartz tuning fork is an excellent basis for the heart of an atomic force microscope, the force sensor.

Christoph Gerber introduced me to Jochen Mannhart, who had worked at the IBM laboratory in Rüschlikon and had taken over a chair in experimental physics at the University of Augsburg. Jochen Mannhart offered me a position and so, starting in January 1997, I had enough freedom to find out in the lab whether the qPlus sensor was actually superior to the silicon cantilever.

This opportunity was a stroke of luck—although years earlier the "experience-diluted university" seemed rather unappealing to me, it was now the perfect environment to try out and optimize this still immature idea. Therefore, it is also important to preserve the freedom of research at the university and not to weave the universities into the supply chains of industry as recently emphasized in an FAZ article by Ferdinand Evers [20].

Already the first experiments were promising, and in December 1999 the subatomically resolved atoms appeared.

IV. The atom as the elementary building block of the world

"If, in some cataclysm, all of scientific knowledge were to be destroyed, and only one sentence passed on to the next generations of creatures, what statement would contain the most information in the fewest words? I believe it is the *atomic hypothesis* (or the atomic *fact*, or whatever you wish to call it) that *all things are made of atoms—little particles that move around in perpetual motion, attracting each other when they are a little distance apart, but repelling upon being squeezed into one another*. In that one sentence, you will see, there is an *enormous* amount of information about the world, if just a little imagination and thinking are applied."

With this observation, the eminent American physicist Richard Phillips Feynman begins his legendary three-volume textbook, *"Lectures on Physics"* [21].

The universe, including our solar system and the earth, consists essentially of atoms, which differ in the number of their protons and neutrons. The number of protons determines the chemical element, the number of electrons usually automatically adjusts to equal the number of protons to yield an electrically neutral atom.

The atomic hypothesis goes back to antiquity. Democritus of Abdera, born in 460 or 459 B.C., was one of the main proponents of atomism, according to which matter consists of minute particles that cannot be further divided. The English physicist John Dalton (1766–1864) established the law of multiple proportions and thus provided strong arguments for the validity of the atom hypothesis, which had previously been postulated only speculatively. Further evidence for the validity of the atomic hypothesis came from statistical thermodynamics, for example from the work of the Austrian physicist and philosopher Ludwig Eduard Boltzmann (1844–1906).

The first experimental images of atoms were achieved by the field ion microscope of Erwin Müller, but its applications were limited because it was only possible to observe the atoms on a sharp metal tip.

The atomic hypothesis is a fascinating idea that captivated even the Dalai Lama and inspired him to write his book *"The World in a Single Atom"* [22]. Because of their tininess, atoms are difficult to see: Fig. 7 provides a size comparison between an atom and a human.

The smallest and lightest atom is the chemical element hydrogen, consisting of an electrically positively charged proton and a negatively charged electron. The carbon atom shown schematically in Figure 8 consists of 6 protons, 6 neutrons and 6 electrons.

The mass of the atom is concentrated almost completely on the atomic nucleus, protons and neutrons are about equal in weight, but the mass of the electron is only about 1/2000 times as large as that of a proton.

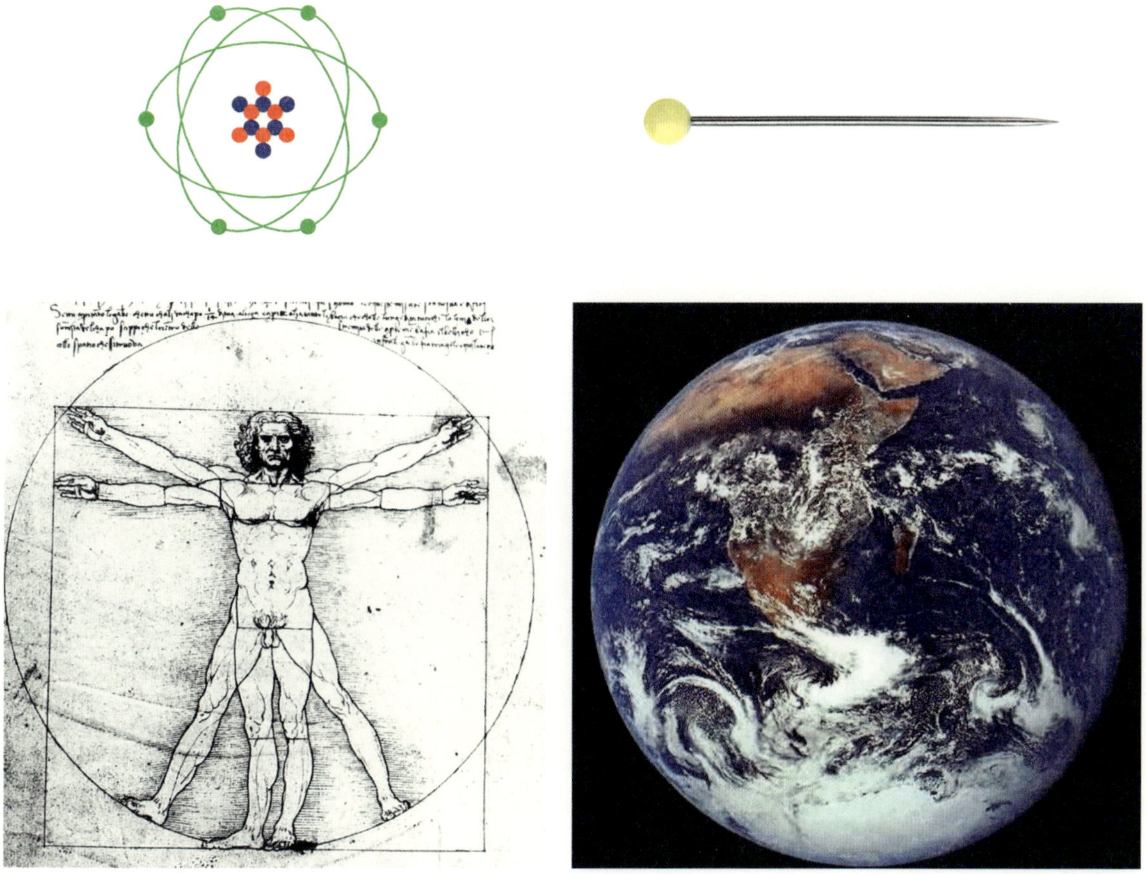

Figure 7. Size of an atom—atoms have diameters between 0.1 and 0.3 nanometers, the size of an atom relates to that of a human being like the size of a pinhead to that of the earth.

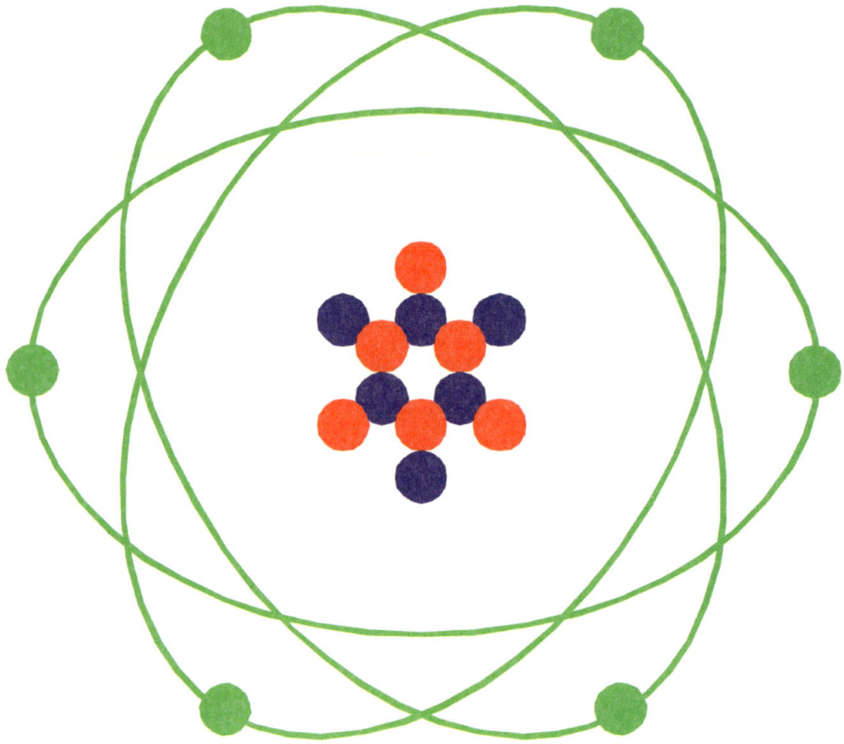

Figure 8. Schematic sketch of a carbon atom. The atom consists of the nucleus with 12 nucleons—6 positively charged protons (red) and 6 electrically neutral neutrons (blue). The nucleus is drawn here greatly enlarged, its diameter is only about 1/100000 of the total atomic diameter. The electron shell consists of 6 electrons (green), which move around the nucleus and are responsible for the chemical properties of the atom.

The forces between the charged particles of protons and electrons, like gravity, are inversely proportional to the square of their distance, but $4 \cdot 10^{42}$ times as large. However, unlike gravity, which is always attractive, these electric forces are repulsive to like charges and attractive to opposite charges. Because of the tremendous electrical forces, the charges usually cancel each other out, and atoms are normally electrically neutral, having just as many electrons as protons.

Consequently, in the nucleus of the atom, the protons repel each other with a tremendous force; the neutrons can be thought of as glue holding the protons together. Light elements, such as the element carbon shown in Figure 8, often have equal numbers of protons and neutrons, while heavy elements have more neutrons than protons.

The chemical element is determined by the number of protons, so the element carbon with the designation $^{12}_{6}\mathrm{C}$ has 6 protons, 6 neutrons and an atomic mass of 12 atomic mass units (protons and neutrons are approximately equal in weight). In nature, however, the radioactive isotope (from ancient Greek *isos* "equal to" and *topos* "location, place"—at the same position in the periodic table of the elements) $^{14}_{6}\mathrm{C}$ also exists.

While the stable isotope $^{12}_{6}\mathrm{C}$ is very old, formed a few hundred million years after the formation of the universe 13.8 billion years ago, the isotope $^{14}_{6}\mathrm{C}$ decays into nitrogen $^{14}_{7}\mathrm{N}$ emitting an electron and an electron antineutrino with a half-life of about 5700 years, which can also be used to determine the age of archaeological organic finds. The exact point in time of the decay of a single $^{14}_{6}\mathrm{C}$ is subject to chance and cannot be predicted. If you have one billion $^{14}_{6}\mathrm{C}$ atoms today, however, 500 million of them will have decayed in 5700 years, hence the term half-life.

A. Atomic force microscopy (AFM)

A microscope is usually understood to be an optical instrument, consisting of an eyepiece, an objective, a tube, and a holding and adjusting device. Its name comes from the Greek words *mikros* (small) and *skopein* (to observe).

Like its predecessor, the scanning tunneling microscope, the atomic force microscope has neither an eyepiece nor an objective. The latter is based on scanning an electrically conductive surface with a sharp, conductive tip. If a voltage is applied between the tip and the sample, a quantum mechanical tunneling current can be detected at a distance of slightly less than one nanometer (one billionth of a meter), which can be used to scan the surface (see Figure 9 A).

The name "tunnel current" comes from the explanation of this current from the quantum mechanical tunneling effect. Classically, in fact, current could flow only when the tip and the sample are in contact, because in the case of a gap, there is a potential wall that the electrons cannot overcome. If this wall is thin enough, for example half a nanometer, then the tunnel effect allows the barrier to be overcome. The number of electrons flowing through the barrier per time depends exponentially on the distance. The current increases tenfold as the distance decreases by a tenth of a nanometer. The diameter of a metal atom is about a quarter of a nanometer. Thus, if only one atom at the tip sticks out 40% of an atomic diameter farther than all the others, ten times the current flows through it. The strong localization of the current thus explains the atomic resolution. The strong dependence of the current on the distance and the monotonous increase with decreasing distance decisively facilitates the control of the tunneling microscope, and thus Binnig, Rohrer, Gerber and Weibel were able to demonstrate atomic resolution on silicon with the tunneling microscope already in 1983 [14].

In atomic force microscopy, things look simpler at first glance—after all, atoms repel each other when you press them together. Thus, one could simply press the cantilever shown in Figure 9 B onto a surface and then scan the surface line by line, as is done on a much larger scale with Braille. However, it took almost ten years after the conception of the atomic force microscope before the silicon surface could be resolved atomically [17]. The reasons for this lie in the difficult physical challenges in the nature of chemical bonding forces—the monotonicity and strong distance dependence given in tunneling microscopy do not apply to the forces. The discussion of these particular difficulties in atomic-resolution force microscopy is beyond the scope of this book; details can be found, for example, in [23, 24].

The paths that the tips of these scanning probe microscopes traverse over the surfaces are finally stitched together in the computer to form an image. These images do not come from an optical imaging process—optical microscopes use light as the scanning medium and have a resolving power of only about a thousand atomic diameters. Therefore, some authors consider the designation of these nanotechnological recordings as images to be incorrect.

In my view, the data produced by scanning probe microscopes are indeed images. As pioneers of scanning probe microscopy, we built the instruments ourselves, soldered together the electronic circuits to control the scanning tip, glued the measuring heads together with our own hands. The function of these instruments was far more familiar to us than our understanding of human vision. We considered these self-built instruments, profoundly understood in their function, as an extension of our human senses. When I first succeeded in atomically resolving the silicon surface with the atomic force microscope in the summer of 1994, the first indication of successful observation of individual atoms was an acoustic one. While scanning the surface, I listened in my headphones to the sound of the scanning tip guided in the scanning pattern. At the same time, I ad-

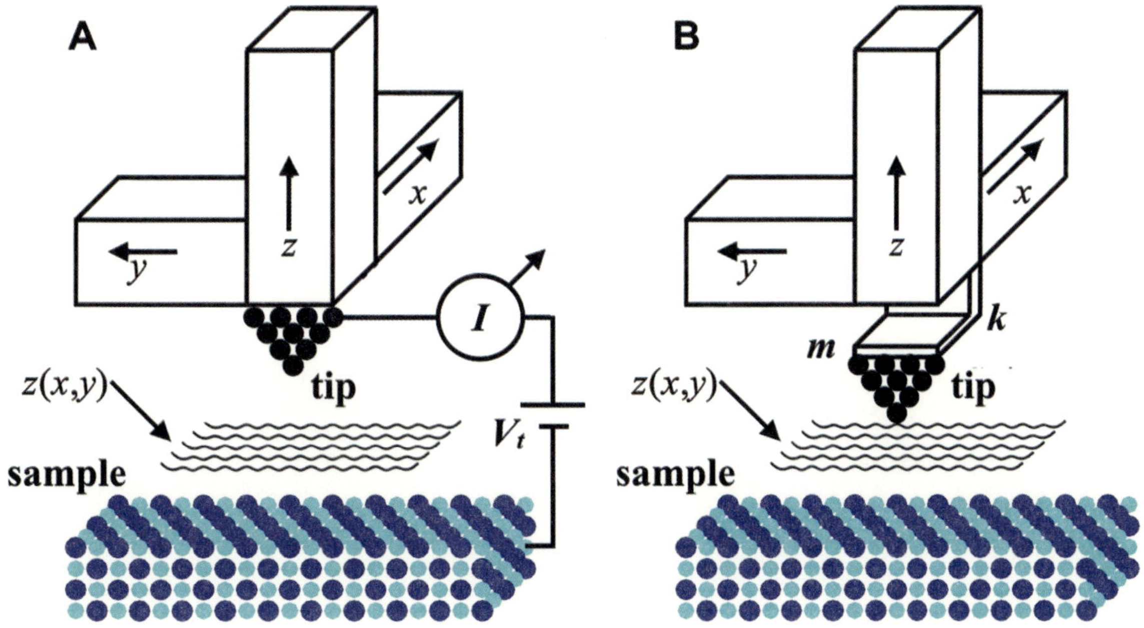

Figure 9. A) Scanning tunneling microscope: a sharp tip is mounted on a piezoelectric tripod, which allows atomically precise positioning and scanning. An electrical voltage V_t is applied between the tip and the sample, resulting in a small electrical tunneling current I_t between the front atom of the tip and the sample atom nearest to it, if the distance is small enough. B) Atomic force microscope: Here, the tip is mounted on a cantilever. Forces between the tip and the sample lead to a bending of the cantilever in quasi-static force microscopy, which has a spring constant k and a mass m. In more precise frequency modulation force microscopy, the cantilever oscillates at its resonant frequency $f_0 = \sqrt{k/m}/2\pi$. When the vibrating cantilever is approacheed to the sample, the frequency changes due to the influence ofthe forces. An image $z(x, y)$ is formed in both types of microscopes by scanning the tip across the sample line by line.

justed the distance of the tip to the surface on a dial and initially heard a "wush-wush" in rhythm with the scanning speed of a few lines per second, similar to the sound of a windshield wiper. With careful further approach, one could gradually perceive through the noise a soft jingling similar to a bicycle bell. This ringing was caused by the individual atoms, which then also appeared on the computer screen as tiny hills.

Later, we learned to measure the force needed to move a single atom on a surface and arrange it as desired. That is why the atoms on the surfaces examined with our microscopes appear as real to us as apples on a table in front of us. The ability to also arrange these atoms as desired and to measure the forces necessary to do so make the image on the computer screen appear to us to be just as real as an image that enters consciousness directly through the eye.

Surprisingly, the surface atoms did not correspond to the previously observed washed-out spheres, but each of the twelve surface atoms within a rhombus appeared split into two humps. These double bumps were identified as internal structures of the imaging atom (see Figure 10).

The explanation of the unusual repeated image of the subatomically resolved tip is provided by Figure 11. The foremost atom of the scanning tip exhibits two crescent-shaped charge clouds, which explain the strange appearance of the surface atoms.

Here perhaps the term "charge cloud" should be explained. For all we know, electrons are at least a hundred thousand times smaller than the

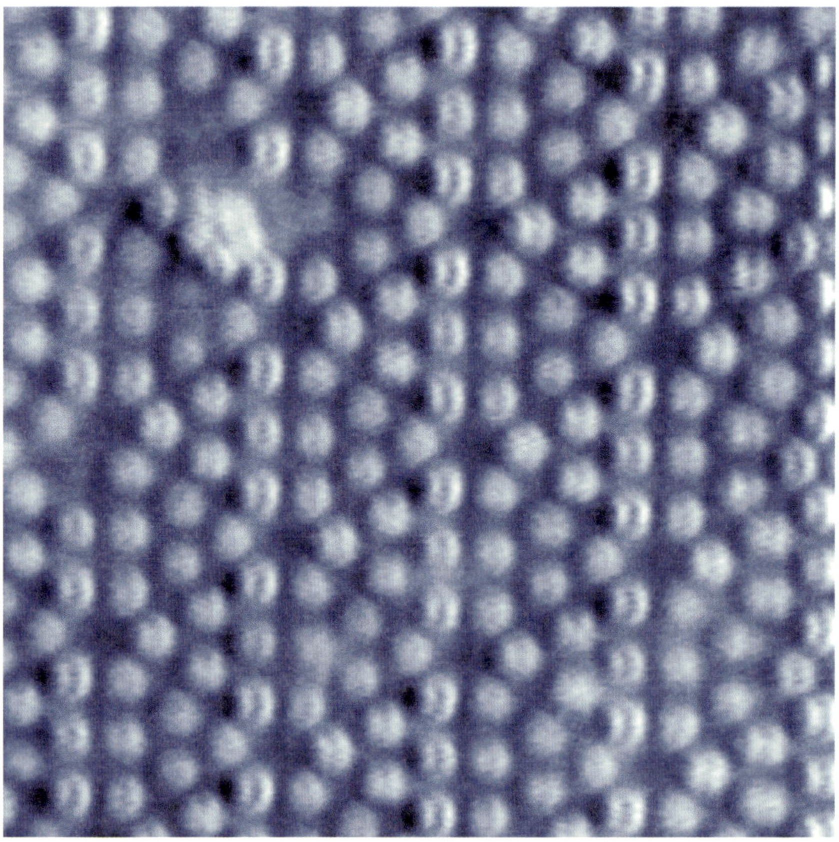

Figure 10. The silicon surface imaged with the quartz tuning fork based "qPlus" force sensor. The surface atoms appear split in half—the explanation is provided by Fig. 11.

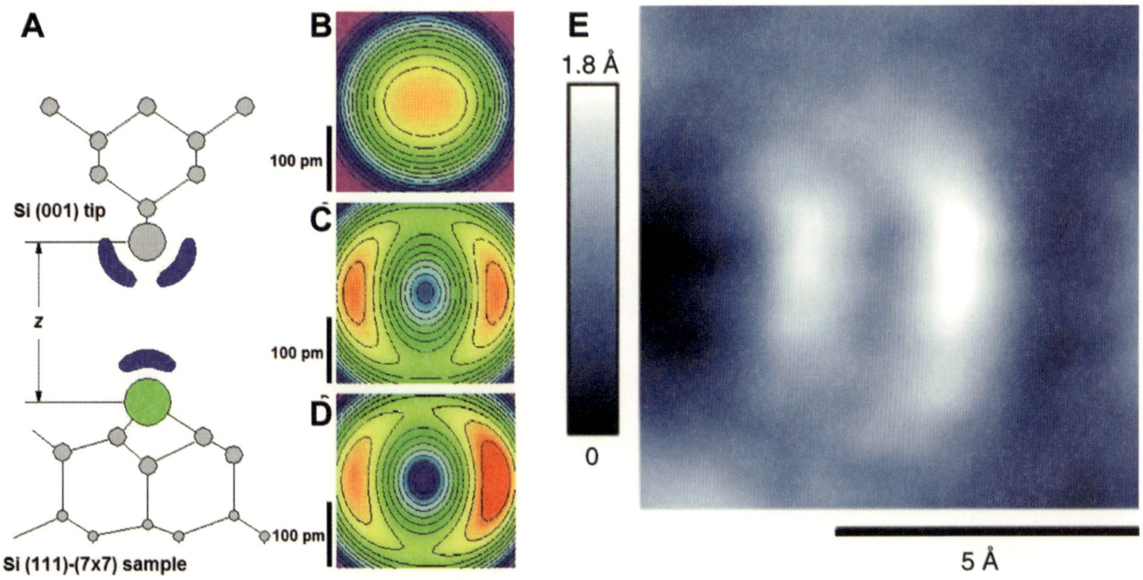

Figure 11. Explanation of the subatomic resolution of silicon atoms. (A) Schematic representation of tip and sample: the front atom of the tip points to the sample with two charge lobes. The top atoms of the sample each have only one charge lobe pointing upward. Calculated images for far (B), nearer distance (C), and tip tilted slightly sideways (D). (E) Experimental image—this was used in the FAZ article.

already very small atoms. Because of the nature of quantum physics, it is impossible to specify the exact location of electrons; they whirl around the nucleus of an atom akin to a swarm of bees around a beehive. So the bright clouds in Fig. 11E are the preferential places where the electrons locate. Since electrons are electrically charged, these places in the image are somewhat inaccurately called charge clouds.

B. What captivated Gerhard Richter about the newspaper article on the atom?

Astonishment, wonder and joy were the feelings when I learned about Richter's "First View". Why did he choose this template? Twenty years later, I still have not found a direct answer to this question.

A letter by Richter (Fig. 12) reads: "Dear Mr. Gießibl, enclosed I am sending you a copy of a text by H. Butin. I don't know what you as an expert think of it, but it pleased me very much, also because I realized that our encounter began 18 years ago and finally resulted in Münster's pendulum. I am grateful to you for that. Cordially, yours, Gerhard Richter."

The laymans expectation to "First view inside an atom", the title of the article in the *Frankfurter Allgemeine Zeitung*, would perhaps be to observe the elliptical orbits often used in symbolic representations of atoms. In fact, the uncertainty principle formulated by quantum physicist Werner Heisenberg in 1929

$$\Delta x \times \Delta p \geq \hbar/2 \qquad (1)$$

prohibits an exact determination of the orbit of electrons. Here Δx means the spatial uncertainty, Δp the momentum uncertainty (the momentum is the product of the mass and the velocity of a particle) and $\hbar = 1.05 \times 10^{-34}$ Js is Planck's constant. Because of the smallness of Planck's constant one does not notice this fuzziness in everyday observations.

In his book *"Der Teil und das Ganze"* [25] Werner Heisenberg discusses the at first unusual transition from the perfectly determined classical physics to the Copenhagen interpretation of quantum mechanics and the increasingly fuzzy linguistic description of physics. However, in the end, even with the language unsuitable for the obscure phenomena of quantum mechanics, an accurate description of reality could be achieved. On page 190 of [25] Heisenberg describes an excursion with Niels Bohr, Carl Friedrich von Weizsäcker, Felix Bloch (all central creators of quantum mechanics) and Bohr's son Christian to a mountain hut near Bayrischzell in Bavaria. Hot and clean water was in short supply at the hut, and so Heisenberg lets Niels Bohr have his say with the statement: "Washing dishes is like using language to describe physics after all. We have dirty water and dirty kitchen towels, and nevertheless it succeeds to clean the plates and glasses finally.

In the same way, our language has unclear terms and a logic limited in its range of application in an unknown way, and still it is able to bring clarity into our understanding of nature with it."

Due to the small mass of the electron ($m_e = 9.11 \times 10^{-31}$ kg) and the tininess of atoms (diameter of a metal atom is about $D_{Atom} \approx 2.5 \times 10^{-10}$ m), the uncertainty principle plays an important role on atomic dimensions.

A quote from Richter provides at least a hint as to why this blurred image of electron clouds might have interested him (page 60 in [27]): "I can't say anything clearer about reality than my relationship to reality, and that then has something to do with blurriness, uncertainty, fleetingness, partiality, or whatever." A further reference is found on page 539 in [27], where Richter says: "I was fascinated by the motif, because the imaging technique in microscopes has reached a point where you can really see the atom, but you can never see it sharply. And that suits me very well, because there's a limit with that."

Accordingly, many of Richter's images reveal an intentional blurriness that may reveal the essence of the object depicted more clearly than a photographically sharp image.

Although Richter's "First View" is a modest work for his œvre in size and physical labor, it obviously has some significance. In September 2018, I received a letter from Richter (Figure 12) in which he sent me a book chapter by Hubert Butin [26]. There, Butin discusses the influence of this print on Richter's following atomic images. "First View" also found its way into a science novel (p. 221 in [28]).

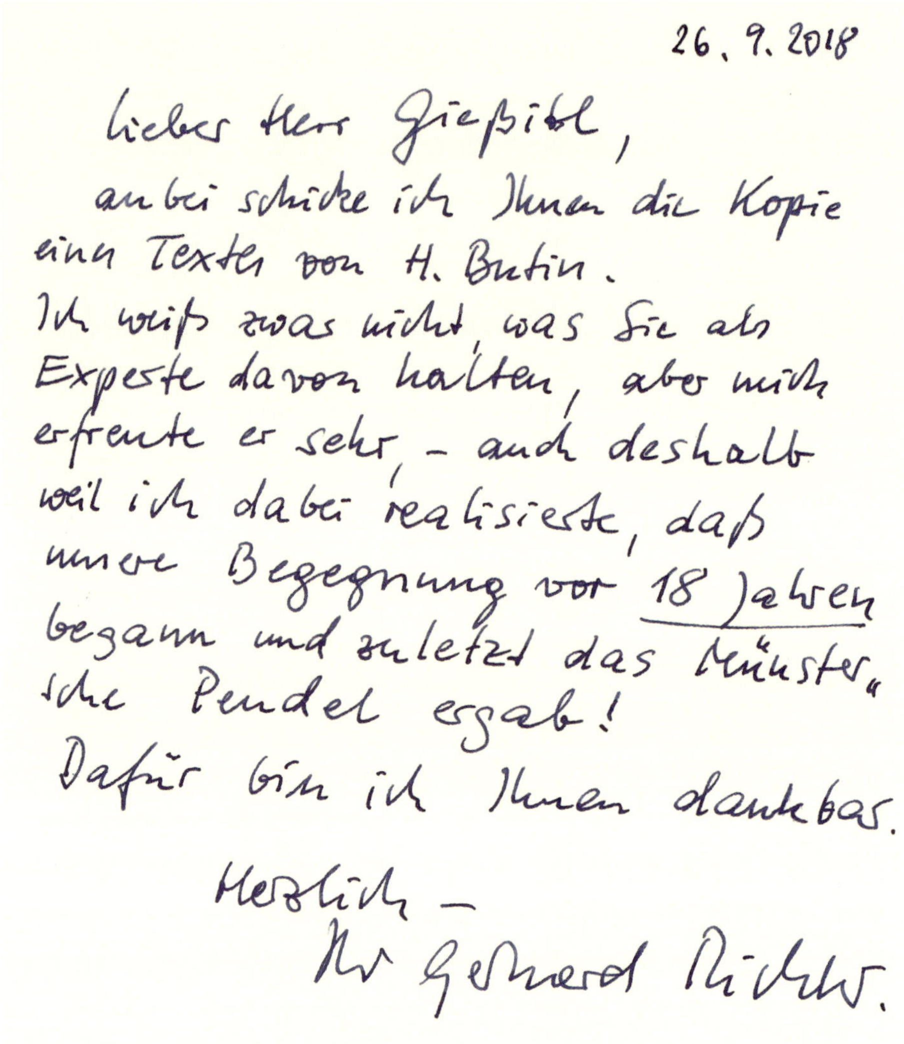

26. 9. 2018

Lieber Herr Gießibl,

anbei schicke ich Ihnen die Kopie eines Textes von H. Butin.
Ich weiß zwar nicht, was Sie als Experte davon halten, aber mich erfreute er sehr, – auch deshalb weil ich dabei realisierte, daß unsere Begegnung vor 18 Jahren begann und zuletzt das „Münster"sche Pendel ergab!
Dafür bin ich Ihnen dankbar.

Herzlich –
Ihr Gerhard Richter.

Figure 12. Accompanying letter by Richter to Gießibl from September 26, 2018, sent along with a book chapter by Hubertus Butin about "First View" [26].

V. Graphite

Since the publication of "First View", I have always kept Gerhard Richter current about our progress in imaging the smallest building blocks of nature. Even if many things were naturally too complicated for him, I wanted to let him at least visually participate in our progress. He reciprocated with invitations to exhibition openings, exhibition catalogs, and books on completed projects such as. *"14 Panes of Glass for Toyoshima, dedicated to futility"* [29] or *"Patterns"* [30], template for the stained glass windows of Tholey, his last major work. He found the reports on our current research interesting and sometimes fascinating, but usually they did not trigger a new work of art by him.

The publication *"Revealing the hidden atom in graphite by low-temperature atomic force microscopy"* [31] was an exception. In this study, we imaged graphite using a combined scanning tunneling and atomic force microscope. In graphite, the carbon atoms form a honeycomb pattern of two-dimensional atomic layers. These layers are stacked such that one atom in each hexagon (drawn in white in Figure 13) lies above an atom in the layer below, while the other (drawn in red in Figure 13) comes to rest directly above the center of a hexagon in the atomic layer below. The scanning tunneling microscope "detects" only the red atoms, while the atomic force microscope also images the previously invisible atom drawn white in Figure 13. Gerhard Richter has inverted the calculated atomic pattern shown in Fig. 13 D, overlaid it with gray paint, and created the offset print "Graphit" (2005) (Figure 14). In his "Atlas" other varieties of "graphite" can be found.

Graphite is essential for artists—it is the central material of a pencil lead. It is also the basis for light and very tensile materials for bicycles, cars and airplanes, it serves as a lubricant. Moreover, carbon is the central chemical element of life, all organic compounds have a framework of the very light element carbon with only 6 electrons. Carbon can form a variety of different bonds—tetrahedral in diamond, planar in graphite and benzene, as well as many petroleum components and related hydrocarbons. In carbon dioxide, which is responsible for global warming, there is a linear bond between the central carbon atom and the neighboring oxygen atoms.

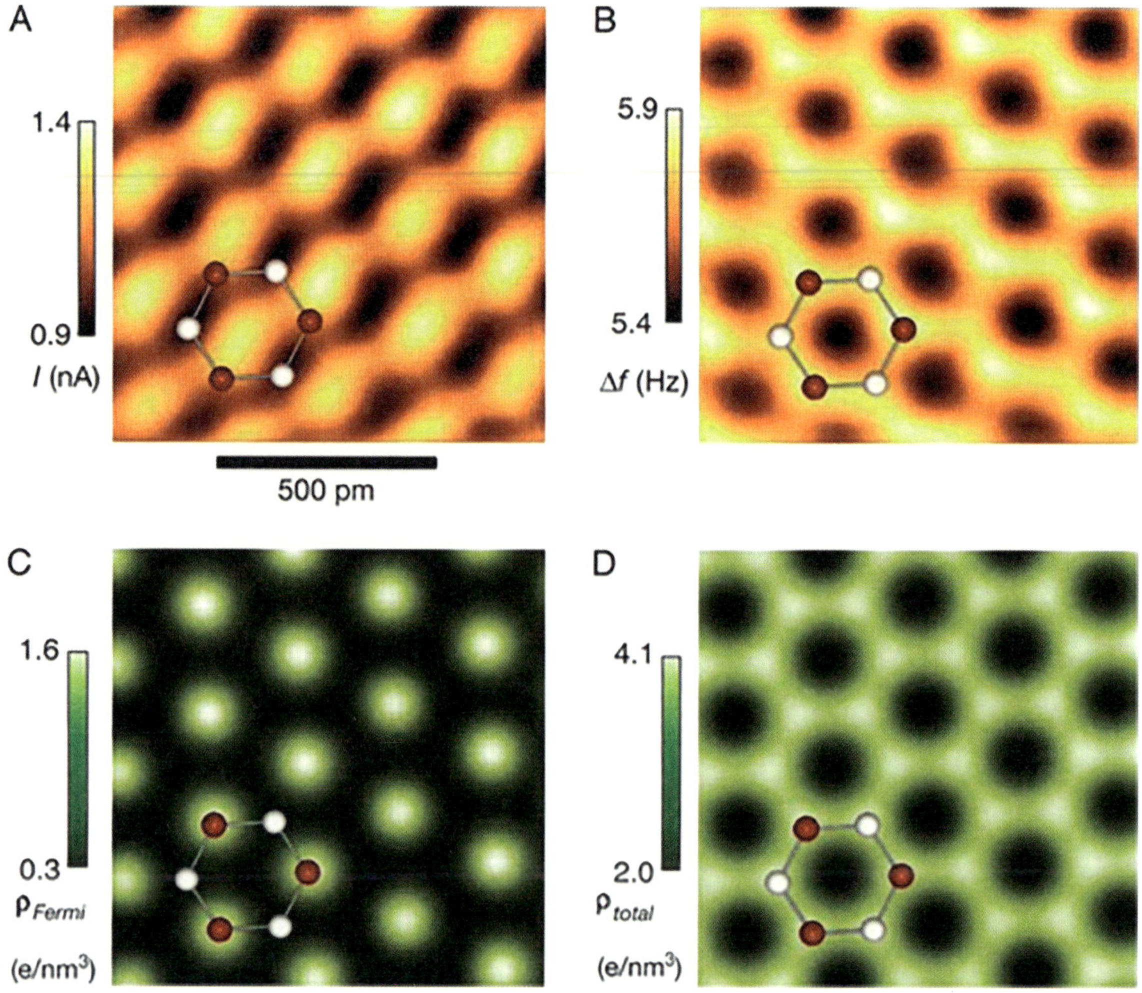

Figure 13. Highly oriented pyrolytic graphite imaged by scanning tunneling microscope (A), by atomic force microscope (B), calculated image for scanning tunneling microscope (C), and calculated image for atomic force microscope (D) [31].

A single atomic layer of graphite is called graphene. Scientists Andre Geim and Konstantin Novoselov had discovered in the early years of the third millennium that single layers of graphene could be created and used to make useful things. They were awarded the Nobel Prize in Physics for this finding in 2010.

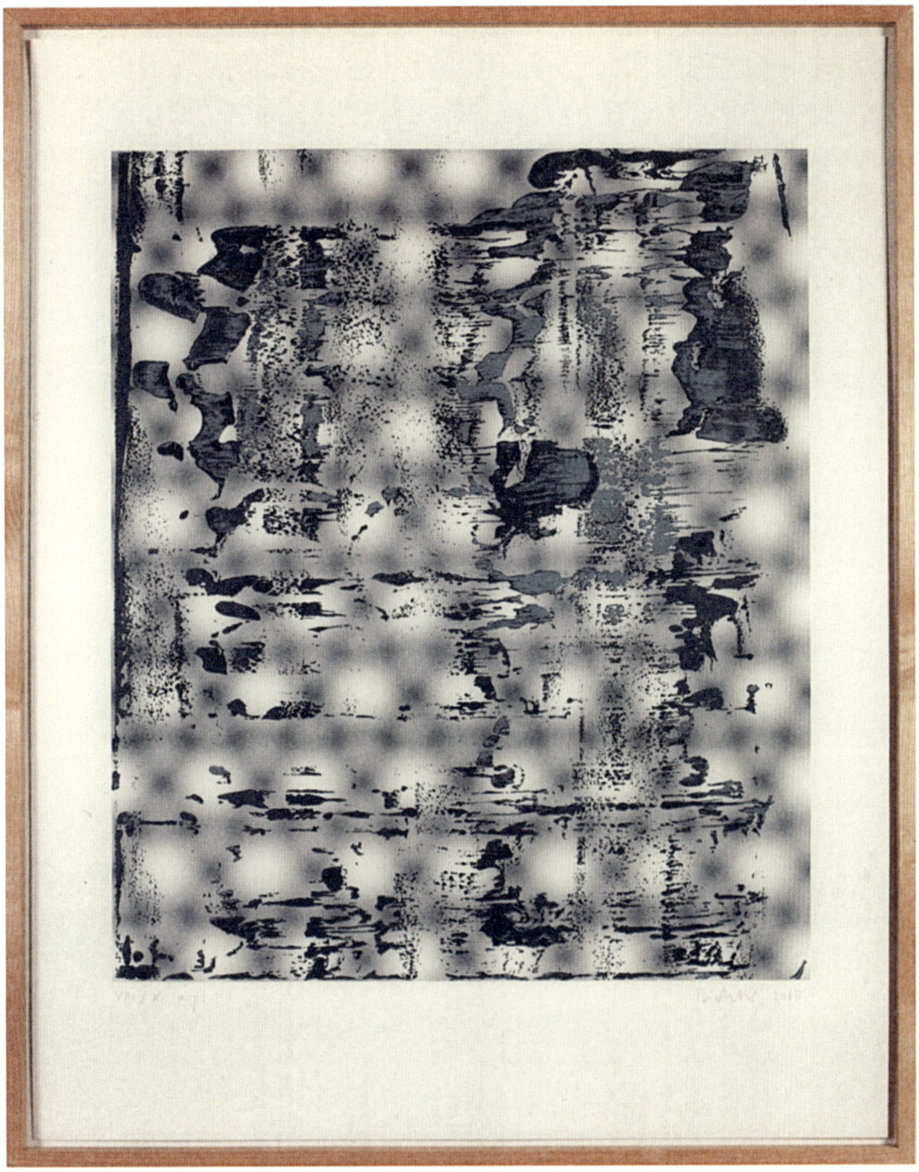

Figure 14. Gerhard Richter "Graphite" (2005), 96 cm × 74.5 cm. Editions-WVZ: 129. Serigraph on duplex offset printing.

VI. Strontium

Around the time of the creation of "Graphite", I visited Gerhard Richter in his Cologne studio in Hahnwald. Among other things, he was working at the time on a commission for the de Young Museum in San Francisco, which was rebuilt in 2005 by the Basel architectural firm Herzog & de Meuron after the old building had been destroyed by the Loma Prieta earthquake in 1989. The task was to create a large mural for the entrance hall, Wilsey Court. "Strontium" measures about 9.10 m × 9.45 m, it consists of 130 individual panels with a size of about 70 cm × 94.5 cm, arranged horizontally in a height of 13 and a width of 10 panels.

Richter is known for his strict quality standards; it is not uncommon for works to be destroyed or painted over if they do not pass muster after having being on private display in the studio for several weeks. On a visit to Hahnwald, he showed me a 2 × 2-arrangement of these "Strontium" panels, hung on a trial basis in the stairwell of his home. "Strontium" is again based on an FAZ article "Atoms in rank and file", which as in "First View" referred to an electron micrograph of strontium titanate [32] published in the journal *Science*.

Strontium titanate has the chemical formula $SrTiO_3$ and crystallizes in the perovskite structure. There, the quadruply positively charged strontium atom sits in the center of each cube-shaped elementary cell, the doubly positively charged titanium atom is located at the eight cube corners, and the doubly negatively charged oxygen atoms are positioned at the six face centers. The material is a transparent insulator, has a refractive index for light similar to that of diamond, and therefore serves as a less expensive substitute for diamond in jewelry. Strontium titanate is popular in science as a substrate for other oxides with interesting properties, such as high-temperature superconductors. Because of its high dielectric constant, it is also being discussed as an insulator for gate electrodes of ultra-small transistors.

The template chosen by Richter is Fig. 1A of [32]; it showed a transmission electron micrograph of $SrTiO_3$ in the [110] crystal direction, where the large bright dots are due to a superposition of strontium and oxygen ions. The small, linearly arranged dots show alternating titanium and oxygen atoms.

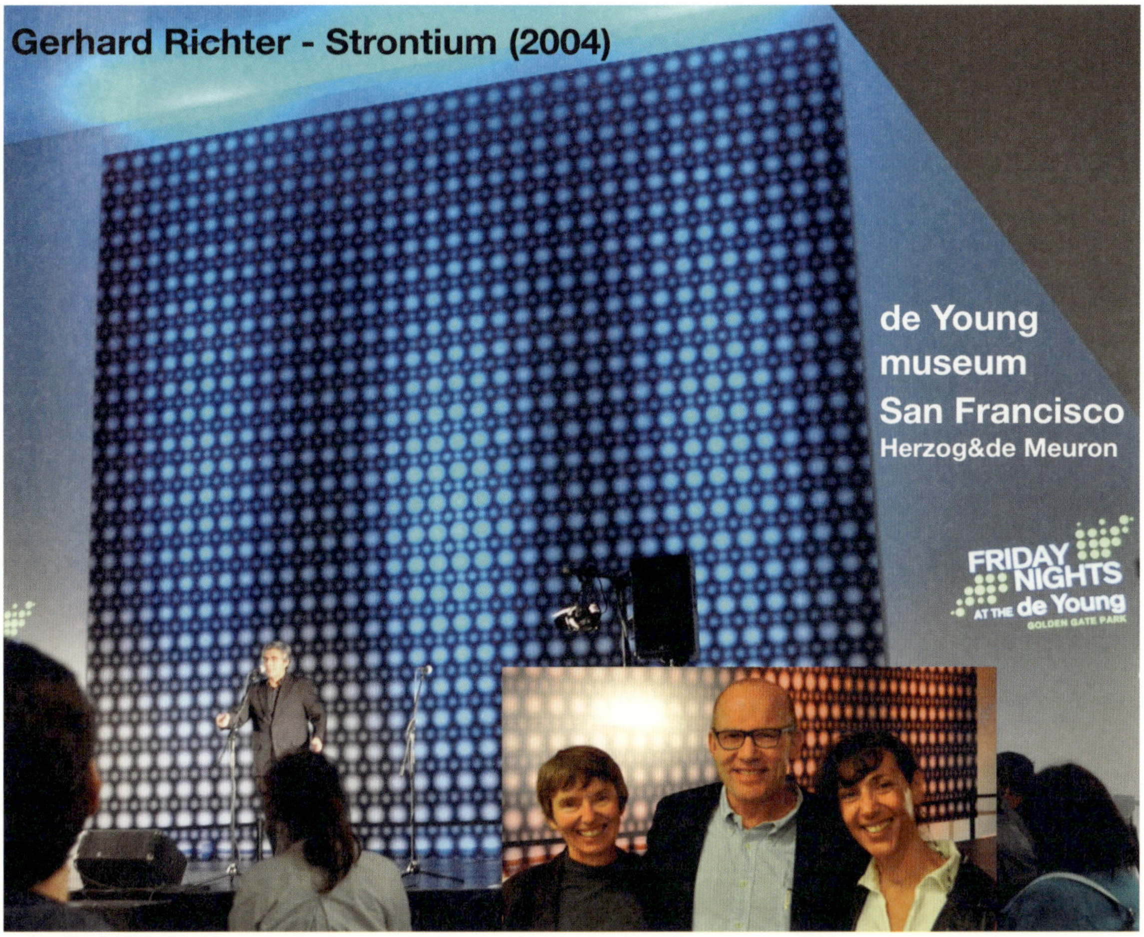

Figure 15. "Strontium" (2004) in the entrance hall of the de Young Museum in San Francisco. Frequently, the image formed the backdrop to the Friday night event "Friday nights at the de Young". The picture shows the author visiting on November 29, 2013, with friends from his earlier days in Silicon Valley.

Silicon Valley and the nearby city of San Francisco have always had a special significance for me—after all, the atomic force microscope was first realized there, not to mention the transistor and the technical achievements derived from it. The energy and dynamism that I was able to feel there during almost three years of work experience have given this mecca of high technology a special place in my heart. This experience had several elements. On the one hand, there was the creative experience of having developed a commercial product there which was exported all over the world. On the other hand, there was an unexpected freedom. The pilot's license acquired at the Stanford Flying Club at the Palo Alto airfield made it possible to explore the magnificent landscapes from the Pacific coast over the Golden

Gate Bridge to the mountains of the Sierra Nevada. Due to the mild climate, it was possible to ride the coastal Santa Cruz Mountains year-round by road bike and motorcycle.

I was delighted to see a work of art by Richter hanging in such a prominent place. In the Wilsey Court of the de Young Museum, "Friday nights at the de Young programs"—celebrated "... after-hours art happenings that include a mix of live music, dance and theater performances, film screenings, panel discussions, lectures, artist demonstrations, hands-on art activities, and exhibition tours." "Strontium" provides an exciting backdrop for these. Figure 15 shows a meeting with friends from my time in California on a Friday evening in 2013 at the de Young in San Francisco.

Sometime in 2013 I remembered a visit to Richter many years before, when he had shown me the proofs of "Strontium" in his house. The specific occasion was the death of Heinrich Rohrer on May 16, 2013, who had been a fatherly friend and mentor to my doctoral advisor (Binnig: "For me, Heini was father figure, role model, emotional and spiritual teacher, and best friend—all rolled into one. An eminent person, with an incredible sense of humanity and kindness." [33]). I thought if these prints were still available, I would like to buy them, give one to Gerd Binnig as a consolation for Heinrich Rohrer's death, give one to each of my two sons as an inspiration to put up in their rooms when they are old enough, and keep one for myself.

I called Richter and asked him if these test prints were available for purchase. He replied in the negative, saying they were hanging in a museum, but that he could check to see if his contract with the de Young Museum would allow more to be made. At first I turned down this generous offer because I did not want to cause any inconvenience. However, during a long run along the Danube the next morning, I couldn't get "Strontium" out of my mind, so I wrote a short email to Richter asking if it wouldn't be possible to have more copies made, after all. A few hours later he called me back with the message that he had checked his contract and that he was allowed to have four more plates produced which he would be happy to sign for me.

Just a few weeks later I was able to pick up the artworks, made in Alu-Dibond by Mike Karstens, at Richter's studio (Figure 16). One of them, 90° turned and somewhat reinterpreted, is shown in Figure 17 as an abstract cross.

In addition to "First View", "Graphite" and "Strontium", Gerhard Richter created other paintings with atomic structures, the oil painting series "Silikat" in gray [34].

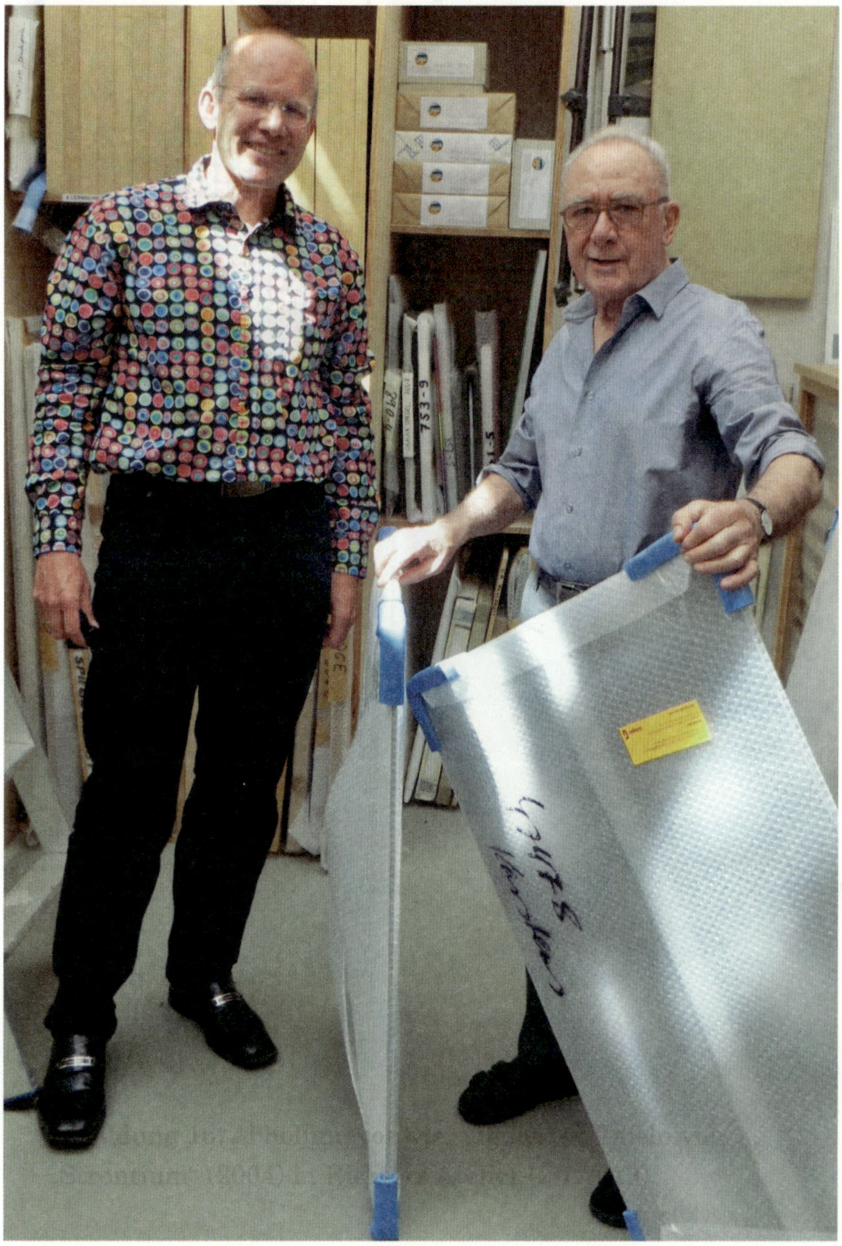

Figure 16. Collection of four signed panels of "Strontium" (2004) in Richter's studio (July 2, 2013).

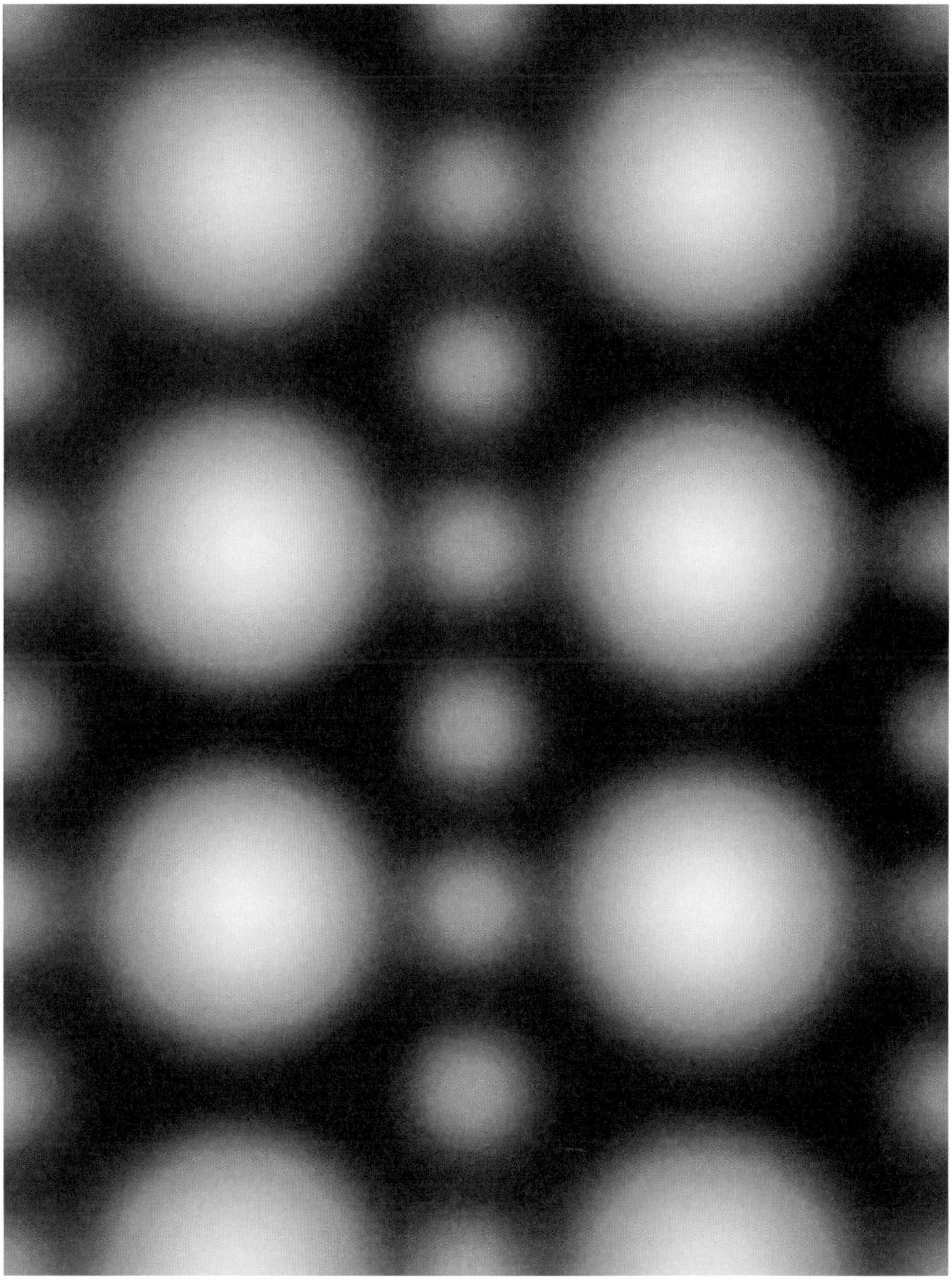

Figure 17. A plate of "Strontium" (2004), rotated by 90°.

VII. Knowledge, doubt and error

The task of science is to explore new intellectual and physical territory. This is shown in Fig. 18. The way of science to advance into new areas of knowledge by exploration of a gray zone is necessarily connected with failures and doubts.

The philosopher of science Karl Popper [35] systematized this process of acquiring knowledge in the following model:

1. Problem
2. Solution attempts ("Theories")
3. Elimination of unsuccessful solution attempts

The scope of a hypothesis depends, of course, on how far it extends beyond existing knowledge. This will be illustrated in Figure 19. A hypothesis that goes only a little beyond what is known will almost certainly be correct, but will make little progress. The greater the distance from the known, the greater the risk of raising an incorrect hypothesis. So it takes courage to fail if you want to make big leaps.

The lack of a linear relationship between the breadth of the advance into the unknown and the quality of the result is reflected in the divergence of quotes from famous people:

Albert Einstein is reported to have said: "Imagination is more important than knowledge, because knowledge is limited." Einstein thus advocates using imagination.

Helmut Schmidt, German chancellor from 1974 to 1982, argues against this: "Whoever has visions should go see a doctor." Schmidt is thus decidedly more cautious and seems to put little value on fantasizing, dreaming, following crazy ideas.

Steve Jobs, founder of Apple and Pixar, in turn, said in a promotional video for Apple. (p. 329 in [36]): "Here's to the crazy ones. The misfits. The rebels. The troublemakers. The round pegs in the square holes. The ones who see things differently. They are not fond of rules. And they have no respect for the status quo. You can quote them, disagree with them, glorify or vilify them, but the one thing you can't do is ignore them because they change things... they push the human race forward, and while some see in them the crazy ones, we see genius, because the ones who are crazy enough to think that they can change the world, are the ones who do."

So Steve Jobs considers breaking rules to be necessary collateral damage in achieving significant progress. However, the reverse is not true, i.e., breaking rules does not automatically ensure progress, but often leads to failure. Rules, after all, are derived from experience, and rules as a condensate of potentially wide areas of experience necessarily must be fuzzy. So it is important to be aware of the precise origin and limitations of rules to identify the domains where the boundaries of these rules should be tested carefully.

The path from the center of knowledge into the edge of not-knowing can also be understood as a creative process, which presumably takes place similarly in science and art. The reprint of the Nobel Prize speech by Binnig and Rohrer [8] shows, in addition to many technical illustrations, a sculpture by the Swiss artist Ruedi Rempfler in Fig. 8 with the remark that art and science are both products of human creativity, and the beauty of nature is reflected by both.

As pointed out in section 3.3, some members of the scientific community had doubts whether we had indeed clearly demonstrated the subatomic resolution capability of atomic force microscopy in 2000. Therefore, we were always on the lookout for new examples, such as the subatomic resolution of a tungsten tip in the 2004 publication [37].

Two aspects of the 2000 [5] and 2004 [37] papers could still be improved:

- First, the subatomic electron structures on the tip were scanned by a periodic grid—in the 2000 paper, the charge lobes of the silicon surface projecting into the vacuum scanned the charge clouds of a silicon atom on the tip of the microscope. In the 2004 paper, the surface atoms of graphite, which according to a recent study [38] may have been locally and temporally converted to diamond due to the pressure of the scanning tip, probed the charge clouds of a tungsten atom on the tip.

- Second, the tip was scanned by the sample because the structures on the sample were "sharper" than those on the tip. Actually, of course, we want the tip of our microscope to scan the sample, not the other way around.

With the offers of a chair in Experimental Physics at the University of Bristol in England and the University of Regensburg and the acceptance

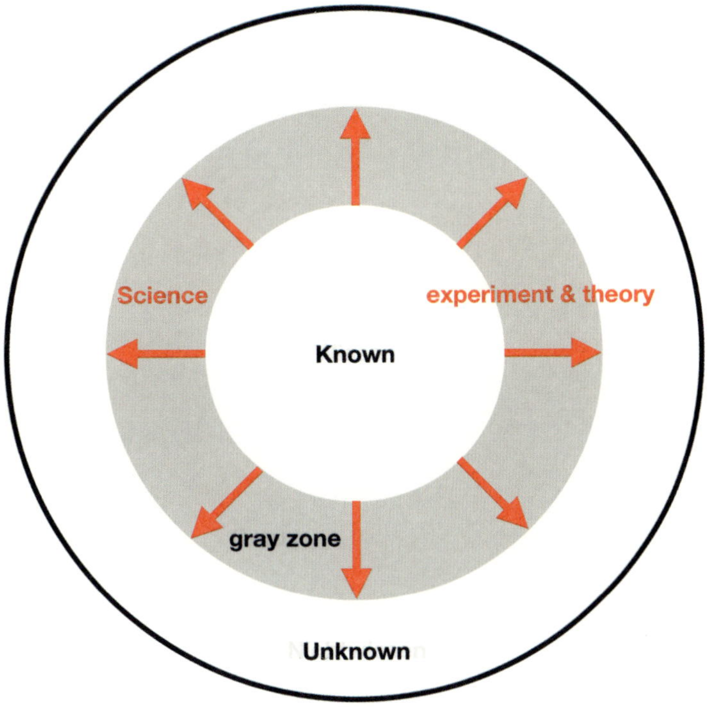

Figure 18. The path of science from not knowing to knowing through exploration of the unknown in a gray area is necessarily fraught with doubts. Of course, these doubts also exist in art.

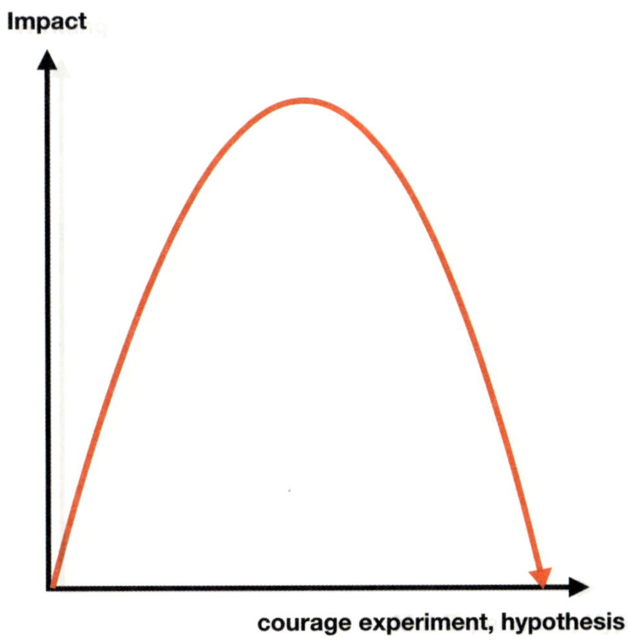

Figure 19. Model description of the scope and influence of a hypothesis as a function of its boldness and courage of imagination.

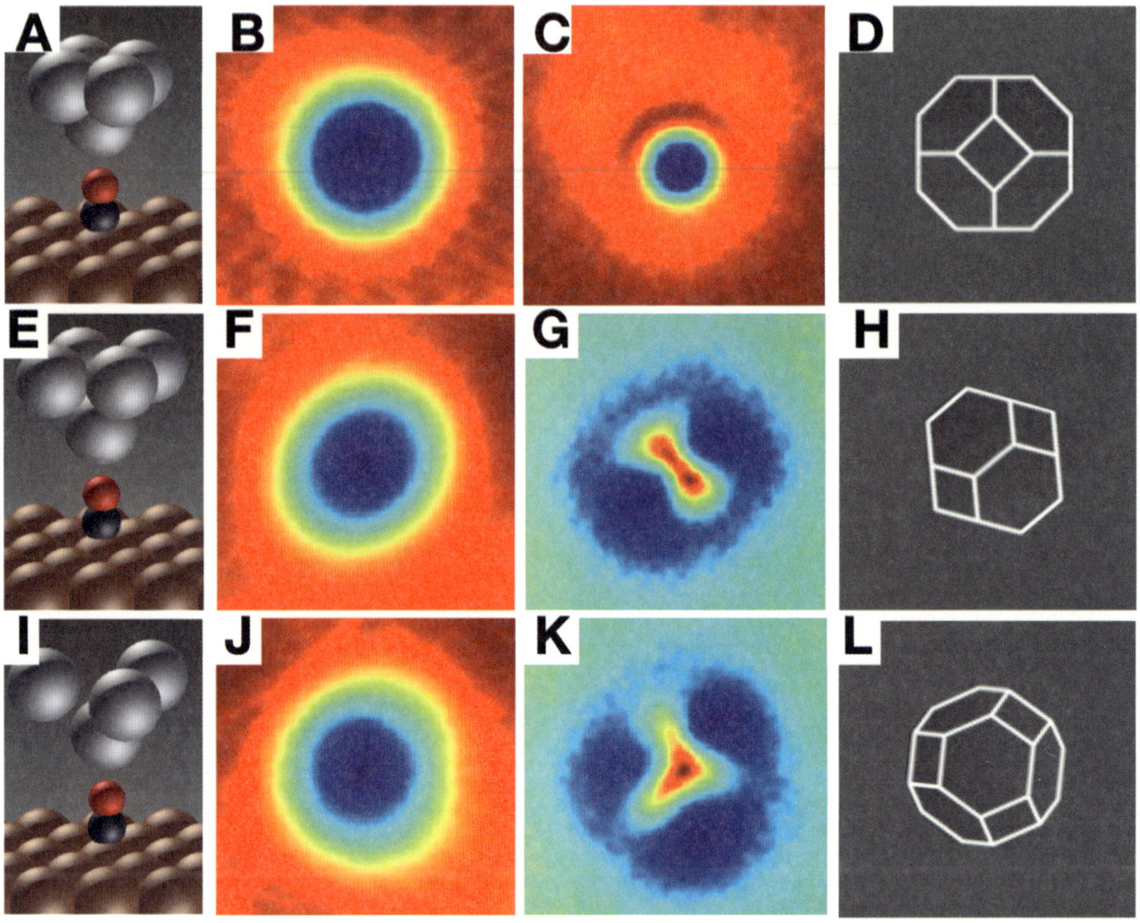

Figure 20. Illustration of a scientific fallacy in the interpretation of experimental measurement data [39]. The measurement tip, of which only the five or four frontmost front atoms are shown in gray in the left column (A,E,I) is scanned by a carbon monoxide molecule (black and red spheres) bound perpendicularly on a copper surface (brown spheres). The bonding of the frontmost tip atom is different for the three cases: In A, the front atom sits on four square-arranged tip atoms, in E on four diamond-arranged tip atoms, and in I on three tip atoms arranged in an equilateral triangle. The second column shows the corresponding tunneling microscopy images B, F, and J. All three results (B,F,J) show a similar size and shape of the tunneling current signal—which is why we originally thought a monoatomic metal tip was present in each of the examples. The force microscopy images in the third column (C,G,K) show much more detail, interpreted as an effect of the bonding ratios of the foremost tip atom to the atomic position behind it (A,E,I). The hypothesis to explain these images lay in the symmetry of the atoms in the tungsten crystal lattice, where each atom occupies a volume bounded by six squares and eight hexagons as shown in D,H,L.

of the latter in 2006, my research opportunities expanded. In the meantime, several manufacturers of high-precision low-temperature atomic force microscopes had incorporated the qPlus sensor into their instruments, and so we were able to quickly pursue more in-depth studies in Regensburg.

In 2012, our team was able to resolve the initial lack of knowledge in the atomic structure of the scanning tip of our microscopy [39] by placing a single carbon monoxide (CO) molecule on a copper surface oriented in the (111) direction. The (111)-orientation results in a surface densely packed with the copper atoms, and the carbon monoxide molecule binds with the carbon atom centered on a copper atom so that it is perpendicular to the surface and the oxygen atom points upward (see Figure 20 A,E,I). The CO molecule is thus a perfect probe to use to probe the scanning tip of the microscope.

This perfect probe led to a very surprising result. In Fig. 20, the images of three different scanning tips are shown, arranged in three rows. The four columns start with the anticipated tip structure to the left, followed by the tunneling microscopy image on the second column, the atomic force microscope image in the third column, and a schematic representation of the tip structure we derived at that time in the right column.

The tunneling microscopy images of the second column showed very similar size and shape for all three tips. Therefore, we initially concluded that all three tips had a single front atom. We attributed the completely different force microscopy imaging to a different crystallographic orientation of the tip—assuming that the atomic layer behind the front atom of the tip conditions the "shape" of the electron structure of the front atom. Following this hypothesis, the front atom of the tip should be on an atomic layer of four atoms arranged in a square in the top row, the middle row on four atoms oriented in a diamond shape, and the bottom row on three atoms in a uniform triangle.

Later (see below) it was to turn out that we had been mistaken. In fact, the tip in the middle row had two foremost atoms, and the tip in the bottom row had three front atoms.

I told my godson Franz Schwarz about the beautiful symmetry of the tungsten atoms and the role of the qPlus sensor in elucidating their structure. On the suggestion of my wife to create a sculpture for the front garden, he gave me the steel sculpture "Wolfram" (see Figure 21) for a round birthday. The body is welded from 3 mm thick steel sheets and shows the Wigner-Seitz elementary cell of tungsten. The shape of this unit cell corresponds to the shape of the atom in the crystal as shown in the right column of Fig. 20.

The swing below symbolizes the quartz tuning fork on which the qPlus sensor is based.

After setting up the sculpture, we experienced a surprise: The body is welded airtight, so when it is warmed by the sun in the morning, the internal pressure rises and the hexagonal surfaces snap outward one after the other with a gong-like sound. In the evening, the sequence reverses and sometimes the last membrane snaps back with a soft sound only at midnight.

The original interpretation of Figure 20 did not stand up to our own subsequent investigations. In 2015, we published a study [40] using the CO terminated tip method found in 2009 by IBM researchers Leo Gross, Gerhard Meyer and colleagues [41]. This involves first creating a monatomic metal tip and then "decorating" it with a CO molecule. Such a tip can produce unexpectedly sharp images because oxygen, the front atom of the tip, amounts to only about 40% of the diameter of a metal atom. It turned out that the image of a single iron atom on copper imaged with this CO tip looked similar to our tip image 20C, while the image of two or three neighboring iron atoms looked like the tip images of image 20G and K, respectively.

Figure 22 now shows the correct interpretation—the CO terminated tip allows the atomic resolved imaging of iron clusters, where the number of atoms can be determined simply by counting. I reported our error to Richter by letter, which he was very pleased about (even physicists can err...). His answer contained a card (Figure 23), in which he marked on an illustration of one of his backglass pictures two differently sized white circular disks with an incised arrow.

Scientific errors have always happened, even to the greatest of our guild from Charles Darwin to Albert Einstein as Mario Livio recently pointed out in his 2013 book *"Brilliant Blunders"* [42].

Another important result of the 2015 [40] study was the clear demonstration of subatomic resolution of the sample, rather than the tip as before. Thus, the second imperfection mentioned above was also resolved. Figure 24 shows the image of a single iron atom on a copper surface. Contrary to expectations, the iron atom does not appear as a simple hill, but as a ring with three local maxima on it. This result could be reproduced arbitrarily because the procedures for its generation were clearly defined. In previous examples of subatomic resolution, there was always a bit of chance involved.

The understanding of the unusual picture of the iron atom as a torus with three hills on it came only some years later. Figure 25 is another example of such an iron atom. In this experiment, published in *Science* [43] in 2019, we

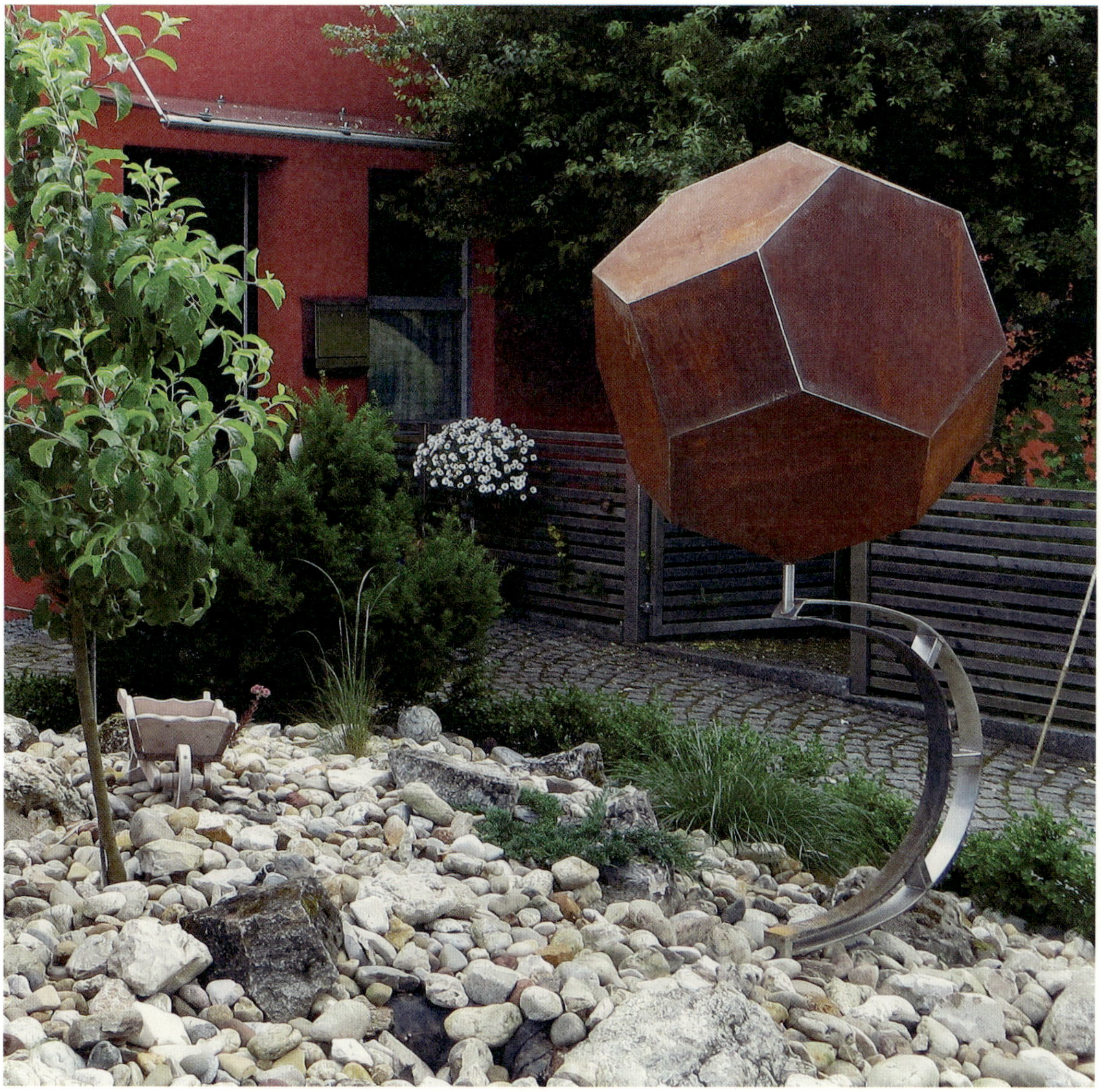

Figure 21. "Wolfram" (Franz Schwarz, 2013). Welded from mild steel and stainless steel, height approx. 1.80 m. The stainless steel (outside) and mild steel (inside) rocker symbolizes the quartz tuning fork of the qPlus sensor, the mild steel body represents the Wigner-Seitz cell of the heavy metal tungsten, which is often used as the material for the scanning tip of scanning tunneling and atomic force microscopes. The Wigner-Seitz cell represents the space occupied by an atom in the crystal structure—periodically juxtaposed Wigner-Seitz cells completely fill the space. Tungsten condenses in the cubic space-centered crystal structure, its Wigner-Seitz cell bounded by six squares on the six side faces of the cube and eight hexagons perpendicular to the four space diagonals of the cube.

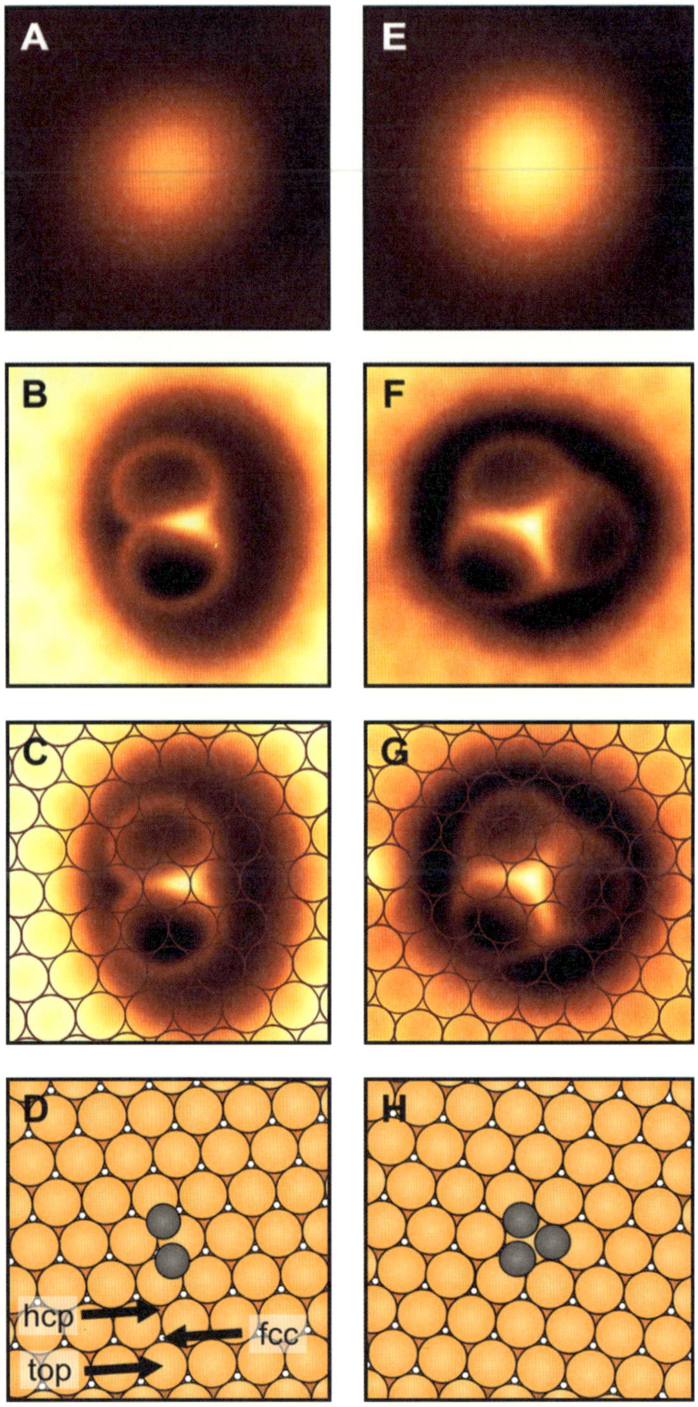

Figure 22. Correcting the misinterpretation from 2012 [39] in 2015 [40]. In this experiment, the carbon monoxide molecule was mounted on the tip of the force microscope. This then imaged two iron atoms (left column) and three iron atoms (right column). The bottom row shows the calculated positions of the iron atoms (gray) on the copper atoms (yellow).

were also able to elucidate the exact origin—it is related to a discovery made by the British physicist Lennard-Jones in 1932, the year Gerhard Richter was born.

In 1932, James Lennard-Jones had postulated that chemical bonding between atoms and surfaces can form a weakly bonded state, called physisorption, and a strongly bonded state, chemisorption. We were able to demonstrate this transition directly experimentally in 2019. The ultimate cause of the subatomically resolved Figures 24 and 25, as well as the earlier subatomic-resolution images, is the strong angular dependence of chemical bonding. This strong directionality of many chemical bonds is also reflected macroscopically, for example in the beautiful symmetry of ice crystals. With the perfectly reproducible imaging of these subatomic structures, this topic is now closed for me. It is to be expected that interesting applications for this precise measurement of the chemical bond will come in the future.

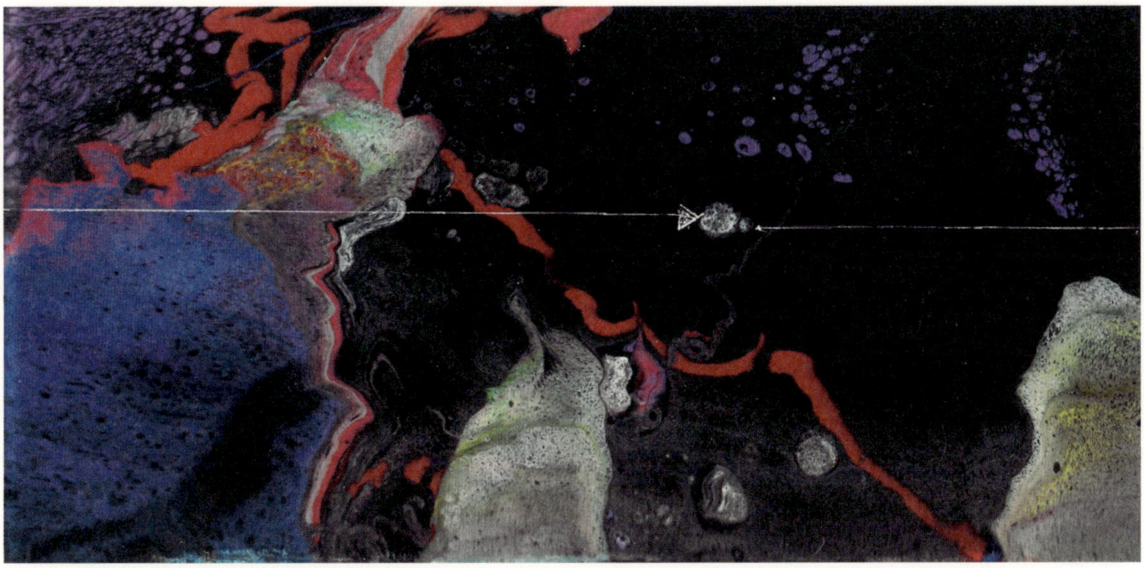

Figure 23. Richter's photo map carved with arrows as a pictorial commentary on our error in the number of atoms present in the images of our tips in Fig. 20.

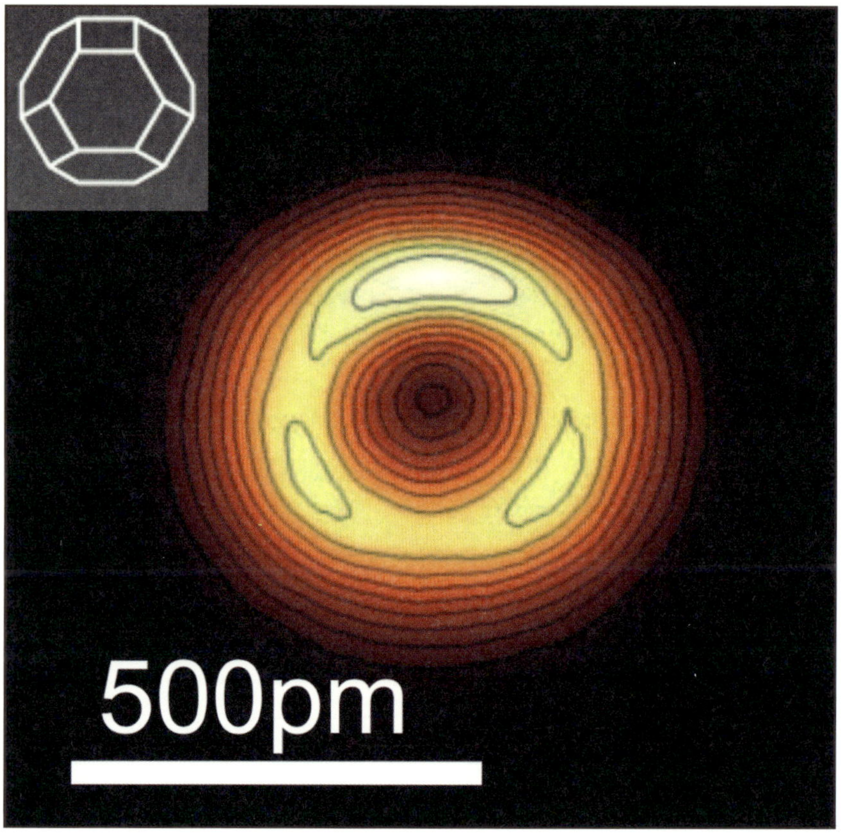

Figure 24. Force microscopy image of a single iron atom scanned with a carbon monoxide tip in 2015 [40]. The schematic "shape" of the iron atom is shown in the upper left corner.

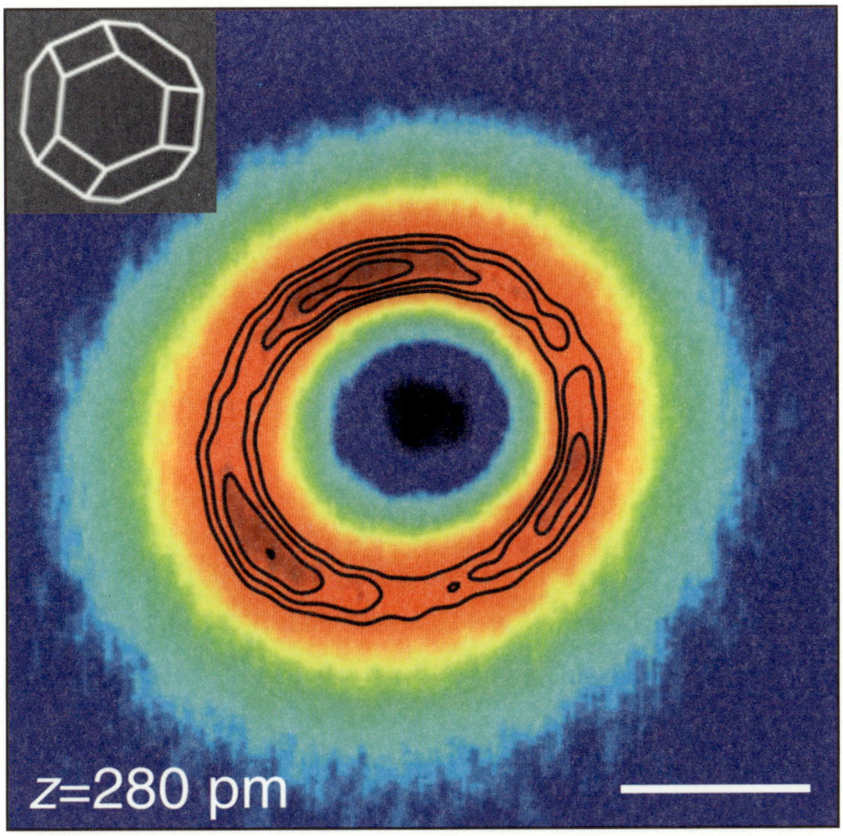

Figure 25. Force microscopy image of a single iron atom scanned with a carbon monoxide tip in 2019 [43]. The image shows the same three cusps as image 24, and the top left corner again shows the schematic "shape" of the iron atom. The slight elliptical distortion is an artifact due to the uncorrected distortion of the scanning unit of the force microscope.

VIII. Foucault's Pendulum

In 2013, Gerhard Richter asked me if I knew how Foucault's pendulum worked and if I could help him build one. I immediately agreed enthusiastically because the pendulum is the elementary form of scientific time measurement and Foucault's pendulum has some analogies to the force sensor in frequency modulation atomic force microscopy.

In both cases, an oscillator is affected by small forces in its oscillation. In the force microscope, the oscillation frequency of a cantilever beam changes due to the influence of tiny forces, and a drive maintains a constant amplitude.

In Foucault's pendulum, minute forces rotate the plane of oscillation over time when the pendulum is suspended in a rotating reference frame such as the earth, while they do not change the oscillation frequency of the pendulum. A drive is also required, which in the simplest case is a human who keeps pushing the pendulum, but of course can also be an automatic drive.

The easiest way to understand this is to think of the pendulum as being suspended from one of the two poles of the earth. The pendulum points to the center of the earth when at rest, and when it is deflected to swing, the earth rotates under the pendulum, so to speak, therefore for an earthbound observer the plane of oscillation rotates by an angle ϕ proportional to the elapsed time, by 360° per day at the north pole and by -360° per day at the south pole.

The evolution of the rotation angle with time as a function of latitude is given by this formula:

$$\phi = t \times 15°/\mathrm{h} \times \sin\theta \qquad (2)$$

At the equator ($\theta = 0$) there is no rotation, at the poles, it varies by ±15° per hour and at the latitude of Münster with $\theta \approx 53°$, where Richter's pendulum was installed, the rotation rate amounts to about 11.8° per hour.

A. Paris

The pendulum bears the name of its inventor Léon Foucault. He had installed the longest of his pendulums in the Panthéon in Paris and it can still be visited today. During a lecture trip to Paris I had the opportunity to view the original shown in Figure 26. This pendulum is approximately 67 m long and is restarted by a museum employee on the hour (see Figure 26). Due to air friction, the amplitude of oscillation decreases to about one third before it is kicked again one hour later. As indicated by the equation above, the oscillation plane rotates 11.5° per hour according to the geographical latitude of Paris.

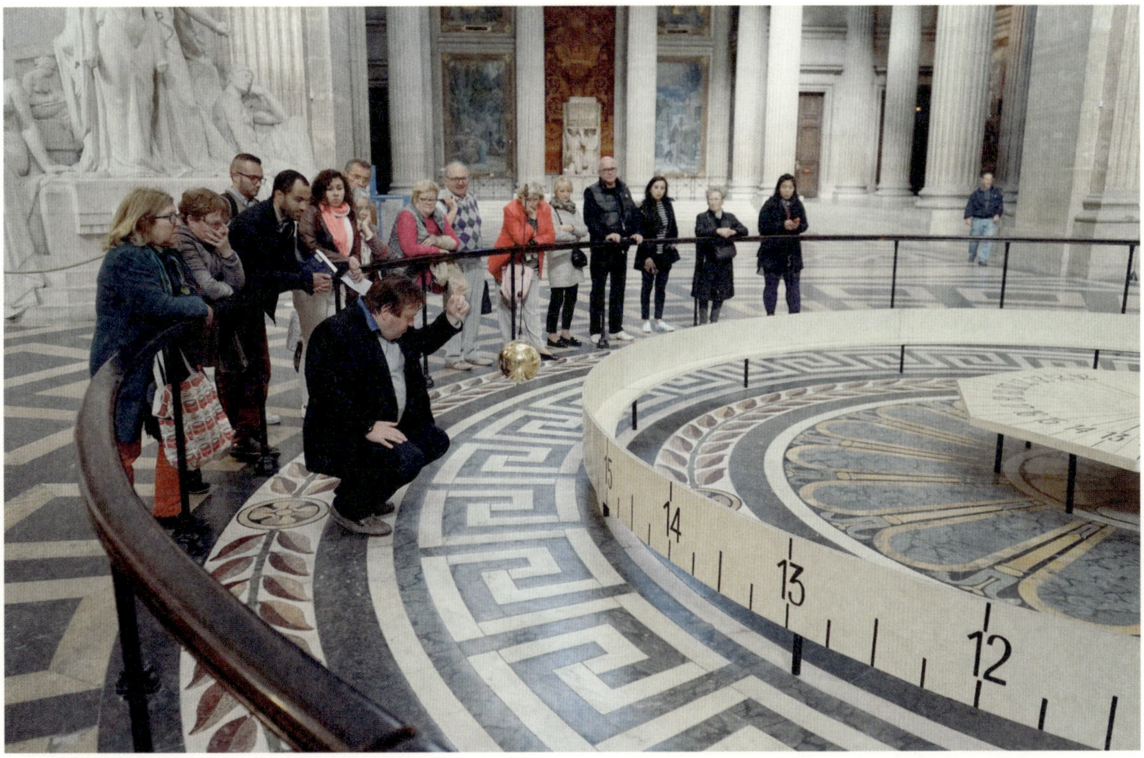

Figure 26. Foucault's pendulum in the Panthéon Paris with a length of 67 meters. The pendulum is struck every hour on the hour during the Panthéon's opening hours by the museum employee standing between the railing and the time scale (shown here at 16:00 on October 6, 2016). After one hour, the amplitude of the pendulum is reduced to about one-third due to air friction. According to the latitude of Paris with 48° 51", the pendulum plane should rotate about 11.5° per hour.

Richter visited the pendulum at the Panthéon during an exhibition of his wife Sabine Moritz also in 2016, but realized while viewing this manually driven pendulum that he clearly prefers an automatic drive.

B. Regensburg

Our first driven pendulum was installed in a laboratory in Regensburg in 2014, shown in Figure 27. This pendulum hangs from a ceiling that is only 3.20 m high. Two years later, we installed a much longer (16 m) pendulum in a stairwell of the physics building, shown in Fig. 28. In order to make the pendulum swing in one plane even after many cycles of oscillation, F. Charron had proposed in 1931 a ring-shaped device [44], which prevents an undesired elliptical motion of the pendulum by making the rope touch a ring near

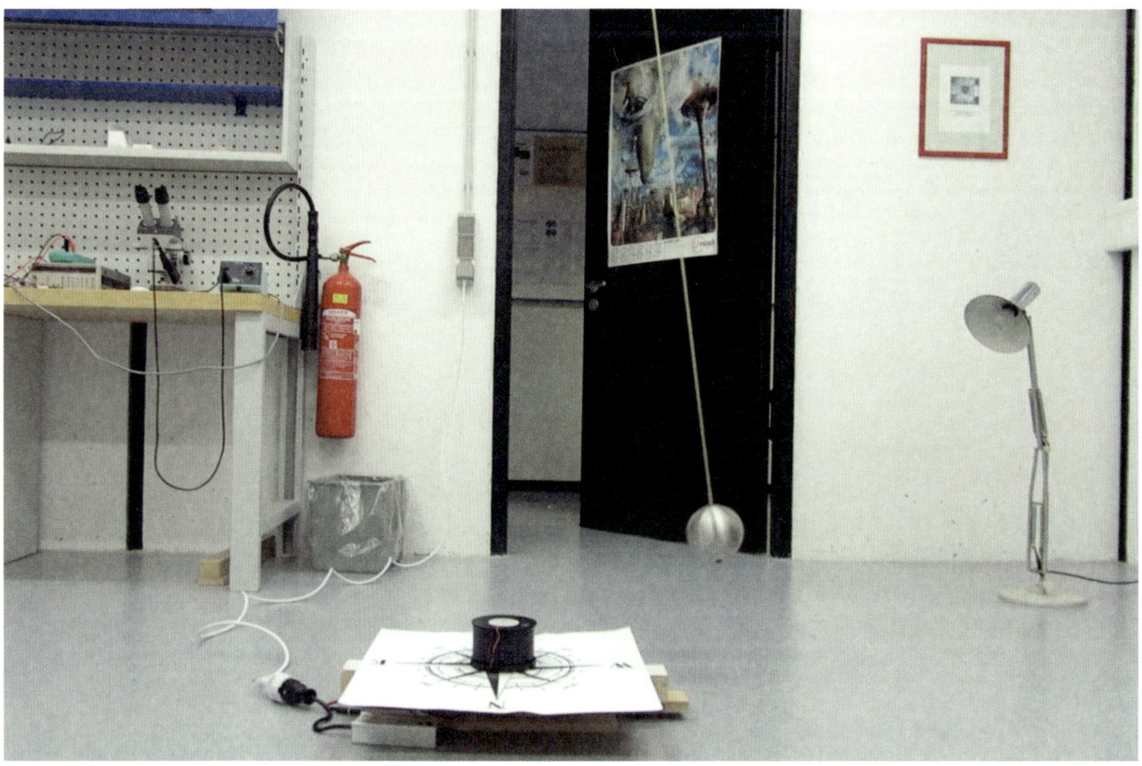

Figure 27. First prototype of Foucault's pendulum at the University of Regensburg with a length of 3.2 m (2014). The latitude of Regensburg is quite exactly 49°, thus one expects a rotation of the pendulum plane of about 11.3° per hour.

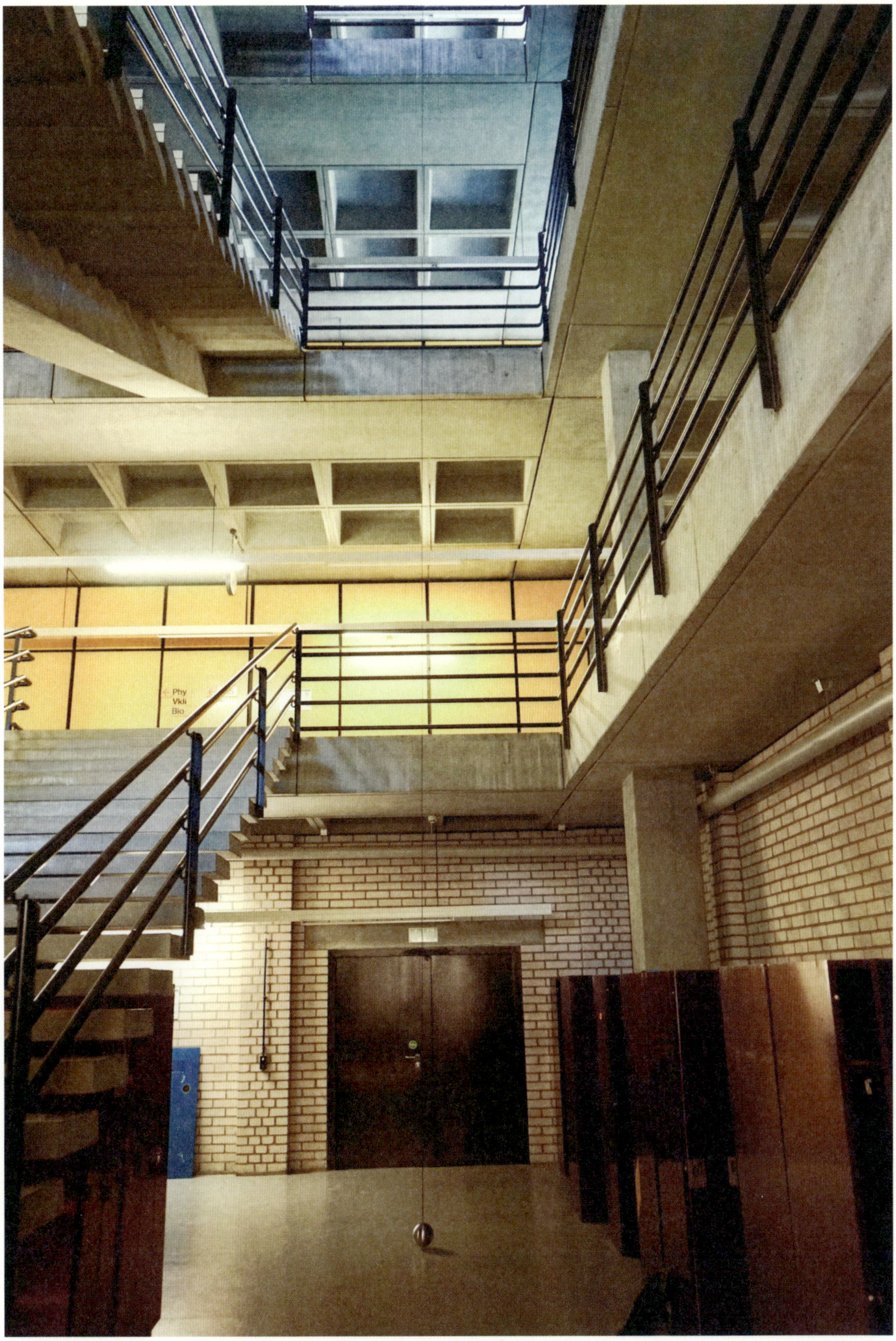

Figure 28. Second prototype of Foucault's pendulum at the University of Regensburg with a length of 16 m (2016).

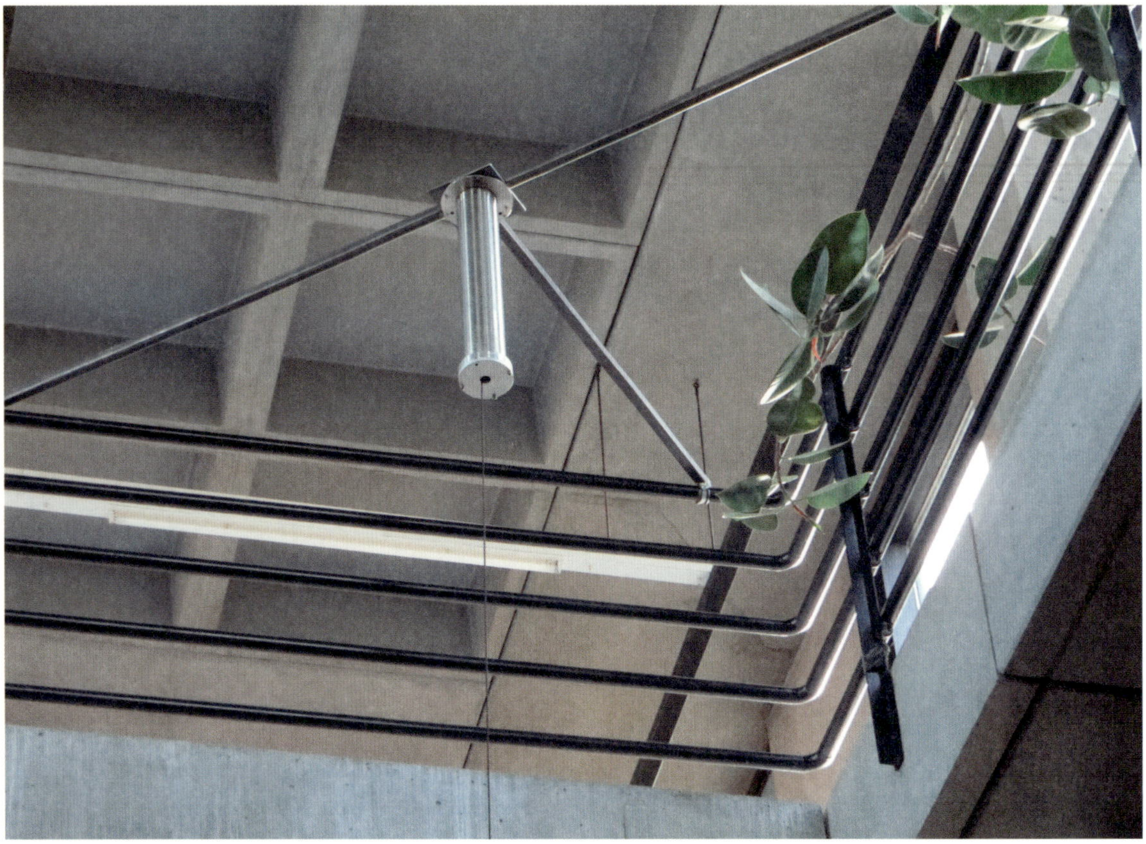

Figure 29. Charron ring of Foucault's pendulum at the University of Regensburg (2016). The Charron ring is only 80 cm away from the clamping point of the rope, i.e. only 5% of the pendulum length.

the reversal point of the pendulum. In an elliptical mode of oscillation, the rope moves sideways just at the reversal point of the pendulum. This sideways movement at the reversal point is effectively damped by friction between the rope and the "Charron ring" sucht that a nice planar oscillation develops. Figure 29 shows such a Charron ring.

C. Münster

Richter was enthusiastic about the pendulum and was now looking for a suitable place to install it. In the planning of the Sculpture Days in Münster, an old abandoned gasometer was suggested. A construction plan of this gasometer with sketched pendulum is shown in Figure 30. This gasometer

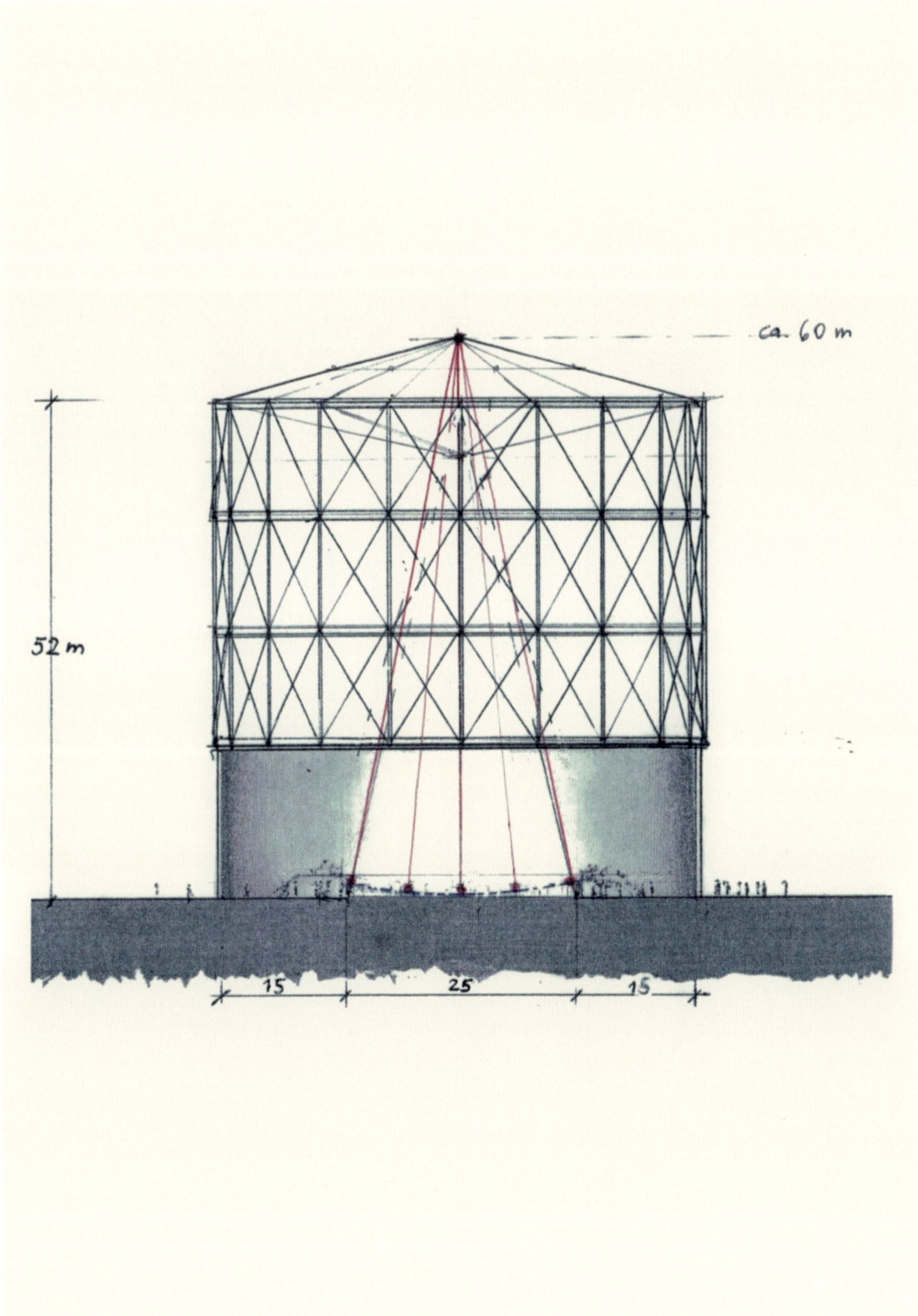

Figure 30. Sketch of the Münster gasometer, for which the installation of Foucault's pendulum was planned (Gerhard Richter 2015).

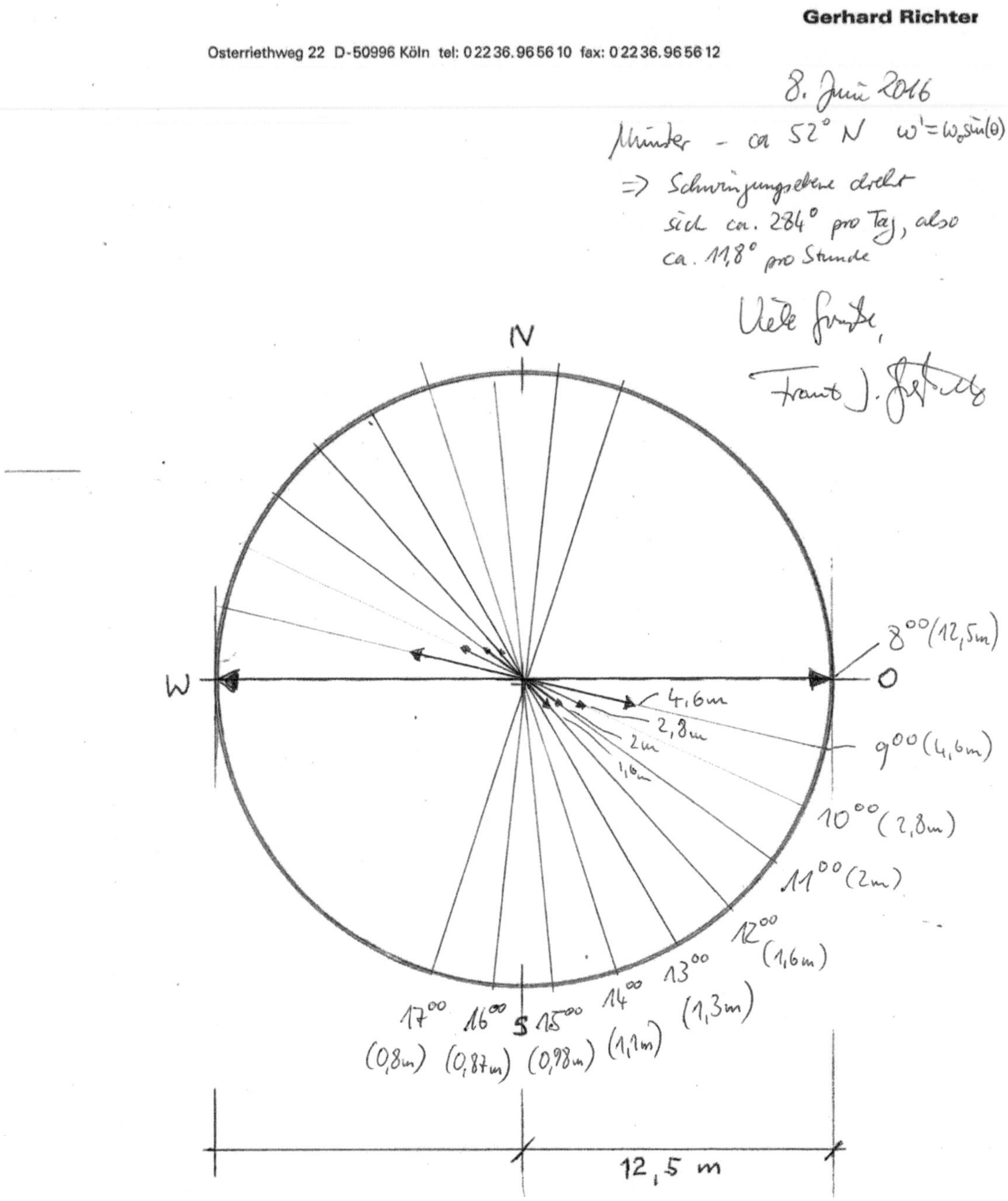

Figure 31. Sketch of the Münster gasometer (top view) with the calculated course of the amplitude over the day if the pendulum is deflected to 12.5 m at 8 am (June 8, 2016).

is exposed to wind and weather, so a reliable electromagnetic drive would be difficult to implement. Figure 31 shows the calculated course of the amplitude of oscillation for a pendulum deflected to 12.5 m at 8 am. At 5 pm the deflection would be only 80 cm if the ball were made of tungsten, a material with a mass density similar to that of gold. This is because the air friction depends only on the diameter of the sphere, and the stored kinetic energy is proportional to the mass. The sphere requested from the Gesellschaft für Wolfram Industrie mbH would have had a mass of 40.4 kg with a diameter of 16 cm. A pendulum with a steel ball would come to rest even faster.

In a sketch from July 2015 (Fig. 32) Gerhard Richter showed me his ideal for an ensemble of pendulum and gray mirrors: a room with space for the pendulum in the center and walls as background for gray mirrors.

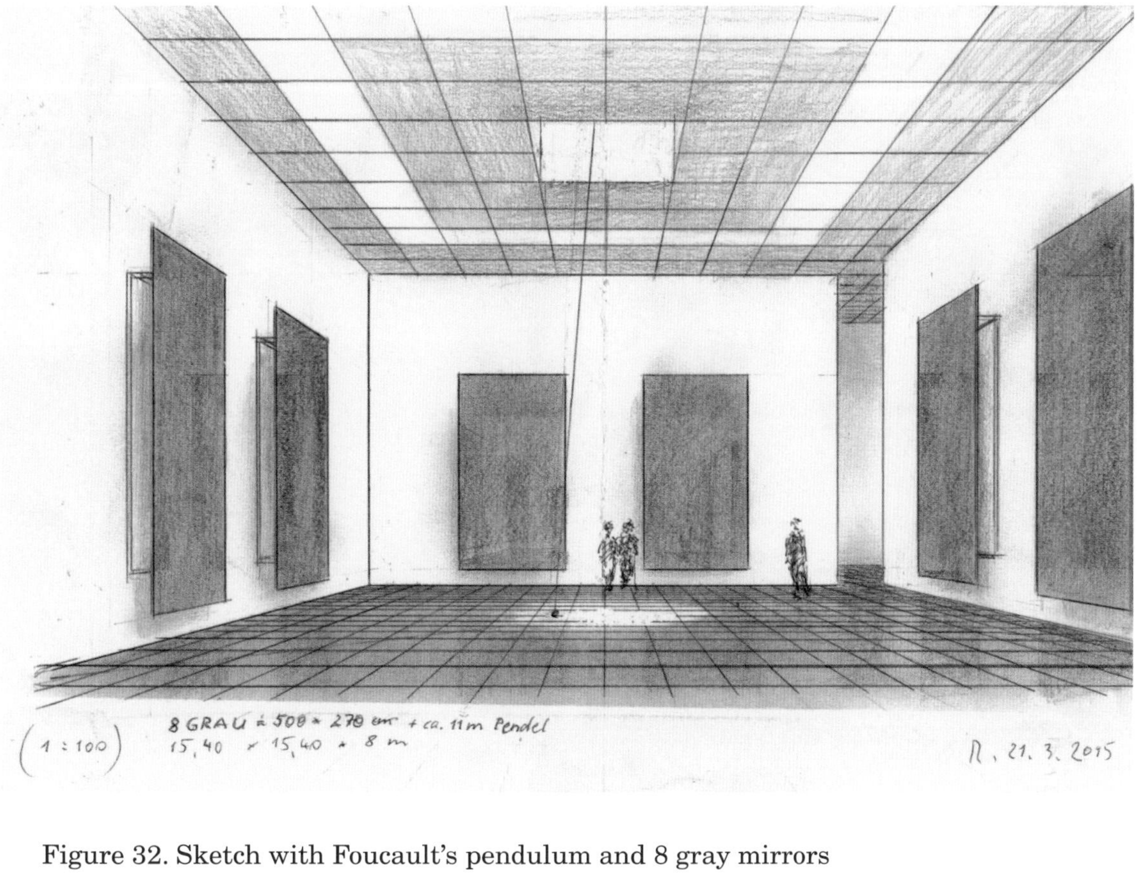

Figure 32. Sketch with Foucault's pendulum and 8 gray mirrors (Gerhard Richter, March 21, 2015).

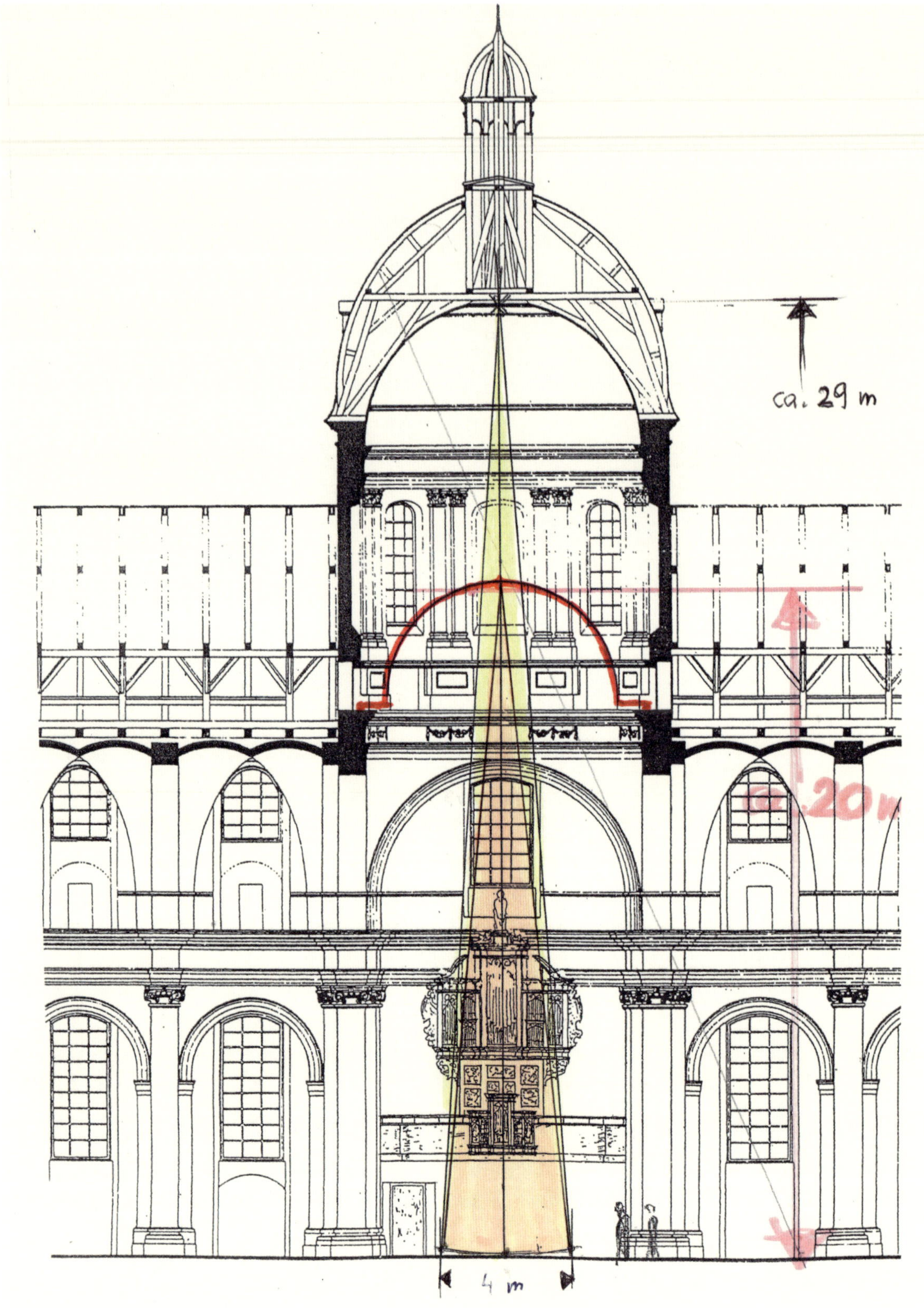

Figure 33. "Two gray double mirrors for a pendulum", Gerhard Richter (2018). Part of one of the two gray double mirrors can still be seen in the upper right. Image from the opening on June 16, 2018 in the Dominikanerkirche Münster.

After visiting the gasometer, however, Mr. Richter was convinced that an open design would not be the right environment, but a far better solution emerged, as he reported in a 2016 email:

From: Gerhard Richter <gerhard.richter@xxx.xxx> 11.07.2016 10:24
To: franz.giessibl@ur.de

Dear Mr. Gießibl,
last week I visited Münster to see the gasometer. A huge, rusty, ruinous frame that may well have its charms, but to me it seemed wholly unsuitable in every way for the installation of a pendulum.
But then the head of culture showed me a beautiful building in the center, the Dominican Church, which has belonged to the city for a few years and which they want to use for cultural purposes, but that doesn't seem to be working out so well yet.
In short, I am very enthusiastic about this basilica and I find it very suitable for a pendulum with a length of about 29 meters and a 2-meter amplitude, which is driven magnetically. There it would have the right protection and a dignified ambience, which could still be appropriately prepared and supplemented. By the way, the mayor is also very interested in this use.
That's just a short note for today to let you know that I will contact you again soon.

Until then, best regards.
Yours,
Gerhard Richter

After that, everything happened very quickly - the city of Münster was eager to move ahead swiftly to realize the project. In the sketch from July 11, 2016 (Figure 33) the rough shape of the pendulum is already shown. The drive for the pendulum in Münster was made by colleagues of the physics faculty of the University of Münster, which makes sense because the local proximity of the builders facilitates and accelerates any necessary maintenance of the pendulum. Only the steel ball and the pendulum rope created in the Regensburg physics workshop were loaned to the colleagues at the University of Münster to test the drive.

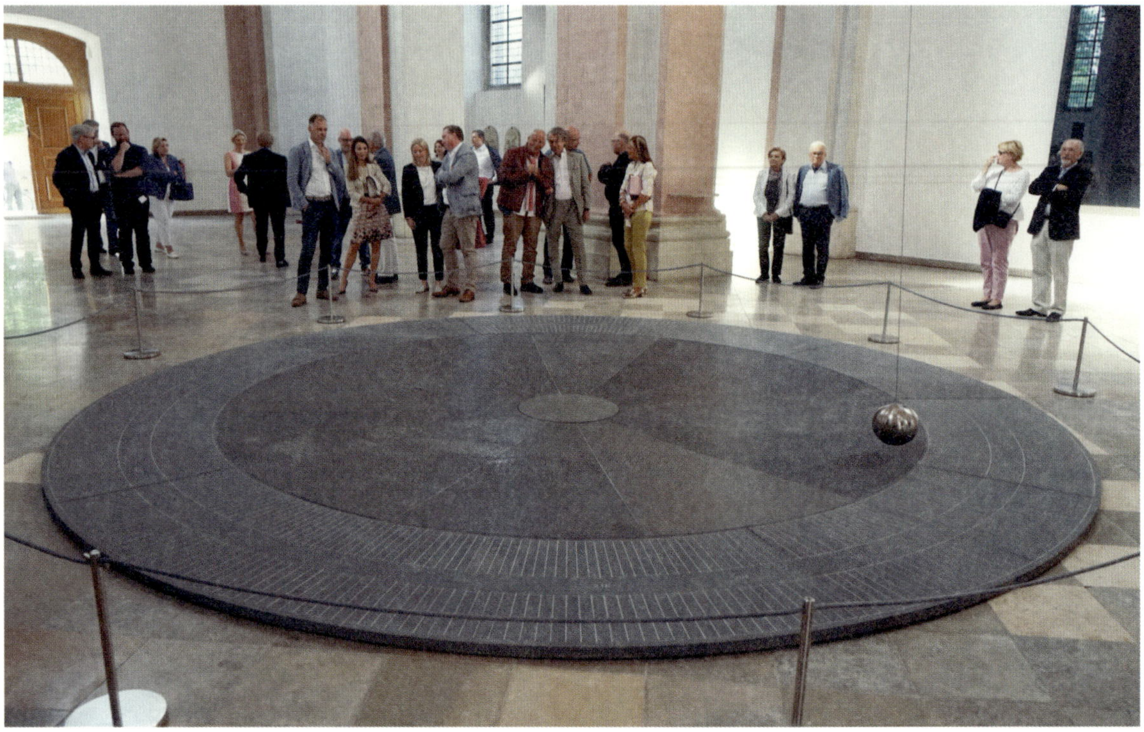

Figure 34. Two gray double mirrors for a pendulum", Gerhard Richter (2018). Part of one of the two gray double mirrors can still be seen in the upper right. Image from the opening on June 16, 2018 in the Dominikanerkirche Münster.

The Münster pendulum was opened as early as June 16, 2018 (Figure 34), and it quickly became a crowd puller. The meditative calm of the pendulum with the gray double mirrors captivates many people.

On the occasion of the opening of the Münster Pendulum, Michael Hagner, Professor of Science Studies at ETH Zurich, published the book *"Foucault's Pendel und Wir. Anlässlich einer Installation von Gerhard Richter"* [45]. There you will find a detailed description of the Münster pendulum as well as interesting historical reflections on Foucault's pendulum.

Even more details about the pendulum can be found in a 2021 book by Dieter Schwarz that focusses on the two gray double mirrors for a pendulum in Münster [46].

IX. Mirror

A. Triangulation

Don Juan, the hero of Córdoba from Max Frisch's comedy "Don Juan or the Love of Geometry" [47] was celebrated for measuring the distance of the cannons to the fortress of Córdoba, and the cannons hit on it, making it possible to capture the fortress. Don Juan's successful determination of the distance was not due to a foolhardy direct tracing of the route to the fortress, but to his knowledge of geometry. Don Juan moved perpendicular to the line connecting the cannons and the fortress until he saw the latter at an angle of 45°, as explained in Figure 35. At an angle of 45°, the distance sought is exactly equal to the distance d' traveled laterally.

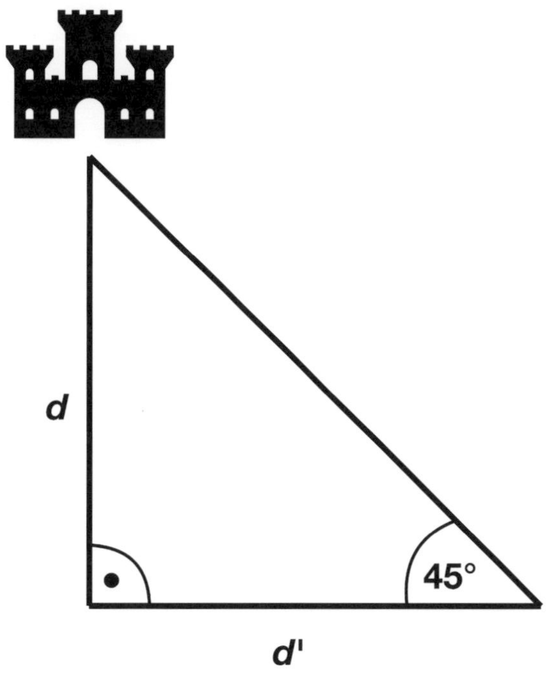

Figure 35. Triangulation: To determine the distance of a target that cannot be reached directly, you can move perpendicular to the line connecting the object and the reference position until you see it from an angle of 45°. The distance d searched for then corresponds exactly to the distance d' covered laterally.

This method is called "triangulation", it is even used in astronomy, where the visual angle of a star is compared at a certain time and six months later. This is because the Earth is laterally displaced by twice the Sun-Earth distance (about 300 million km) during half a year. This results in a measurable angular difference for stars that are not too far from the Earth.

A change of perspective not only allows one to measure distances precisely, but also to better grasp the new, the unknown. Richter's far-reaching experimental approaches to depicting reality and his constant drive to open up new pictorial spaces aroused my curiosity about his approach to reality. To a certain extent, this was obviously mutual, so that decades after "First View" we are still in contact with each other.

In a figurative sense, it can also be called triangulation, when, for example, a painter takes the perspective of a scientist and vice versa, in order to find orientation in a nebulous gray area.

Presumably this desire for a change of perspective has been the driving force behind the many interactions over the last twenty years, and the stimuli in the opposite direction have been so extensive that I can only touch on a few aspects here. Curiosity about the unknown, with its many imponderables and possibilities for error, illustrated in Fig. 18, in conjunction with the sharper view from the sidelines Fig. 35, is the probable reason for the continuing encounters.

B. Beethoven

In a letter from Ludwig van Beethoven to Nikolaus Simrock in Bonn from November 1820, Beethoven exclaims "I would rather write 10000 notes than one letter." [48] Analogously, the communication with Richter was not only through personal encounters, telephone calls or letters, but also through images.

One example is Richter's card from Fig. 23 as an answer to my letter to him with the report about a scientific misinterpretation I had made.

Another example can be found in Fig. 36. The picture was taken in October 2018 at the Regensburg University Hospital, where I was undergoing treatment for an abnormality of the hematopoietic cells in the bone marrow. I asked if I could inspect my own cells in the lab at the university hospital, and photographed them through the eyepiece of an optical microscope using the smartphone camera. I found the picture interesting and sent it to Richter. Richter replied with a signed card shown in Figure 37. The card can be interpreted as an image of cell repair.

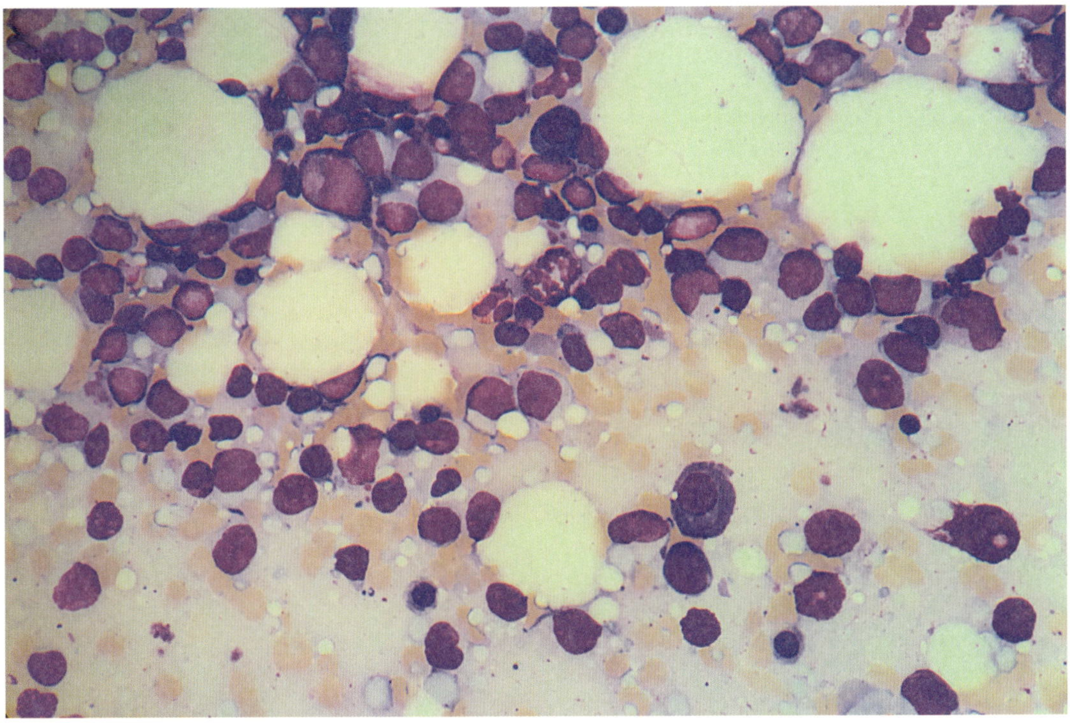

Figure 36. Microscopic image of blood cells, October 22, 2018.

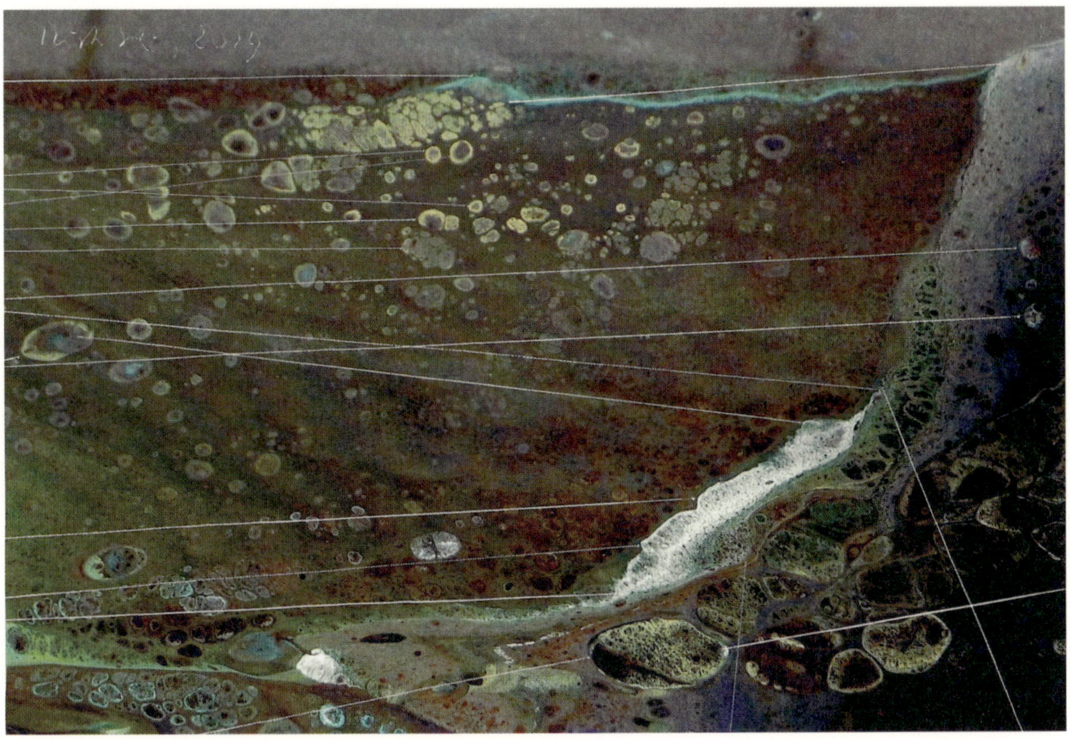

Figure 37. Card from Richter to the author dated January 3, 2019 that can be interpreted as a mechanism of cell repair.

C. Symmetry

Navigating the gray area between knowledge and not-knowing can be supported by models. When we took the images of single iron atoms (Figures 24 and 25), I had a plexiglass model made of an iron atom in its crystal lattice as shown in Figure 38. Of course, atoms do not have corners and edges, the model rather shows the symmetry of the volume available to a single iron atom in the crystal lattice. Iron, like tungsten, for example, crystallizes in cubic-space-centered symmetry and remains in this structure as long as the temperature remains below 910 °C.

In the body-centered cubic structure, one atom sits at the center of a cube and eight atoms sit at the corners of the cube. If we now think of planes which are perpendicular to the connecting lines of two neighboring atoms at their midpoints, the so-called Wigner-Seitz unit cell of the body-centered cubic crystal structure is bounded by eight hexagons and six squares as shown in Figure 38.

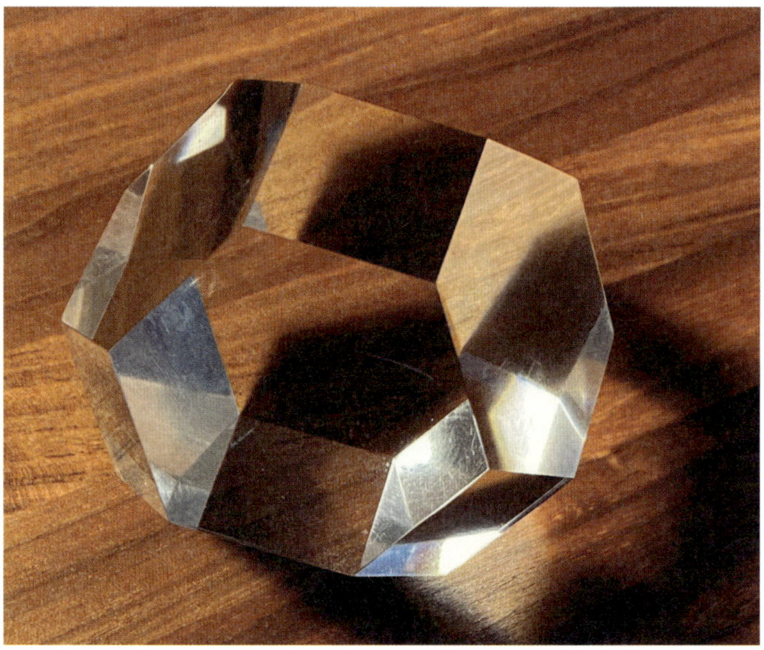

Figure 38. Wigner-Seitz elementary cell of the body-centered-cubic structure of iron and tungsten.

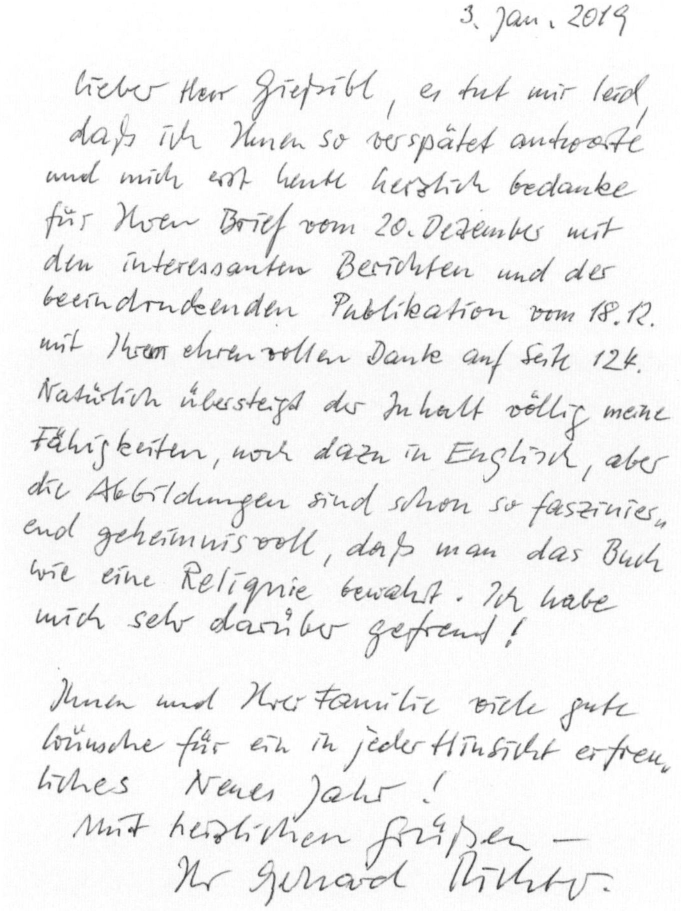

3. Jan. 2019

Lieber Herr Giessibl, es tut mir leid,
daß ich Ihnen so verspätet antworte
und mich erst heute herzlich bedanke
für Ihren Brief vom 20. Dezember mit
den interessanten Berichten und der
beeindruckenden Publikation vom 18.12.
mit Ihrem ehrenvollen Danke auf Seite 124.
Natürlich übersteigt der Inhalt völlig meine
Fähigkeiten, noch dazu in Englisch, aber
die Abbildungen sind schon so faszinierend
und geheimnisvoll, daß man das Buch
wie eine Reliquie bewahrt. Ich habe
mich sehr darüber gefreut!

Ihnen und Ihrer Familie viele gute
Wünsche für ein in jeder Hinsicht erfreu-
liches Neues Jahr!
Mit herzlichen Grüßen —
Ihr Gerhard Richter.

Figure 39. Letter by Gerhard Richter dated
January 3 2019

I sent a copy of this Plexiglas model to Richter because I imagined that he might also find such a model inspiring. In fact, his office manager Mrs. Ell confirmed to me that the model had found a nice place in Richter's studio.

I was very pleased to receive Richter's response (Figure 39) to a review article I had just completed at the end of 2018. It shows that Richter is still fascinated by new scientific territory at an advanced age.

The English translation of the letter reads: "I am sorry that I am so late in replying to you and thanking you very much for your letter of December 20 with the interesting reports and the impressive publication of December 18 with your honorable thanks on page 124. Of course the content is completely beyond my abilities, even more so in English, but the illustrations are so fascinatingly mysterious that one keeps the book like a relic. I was

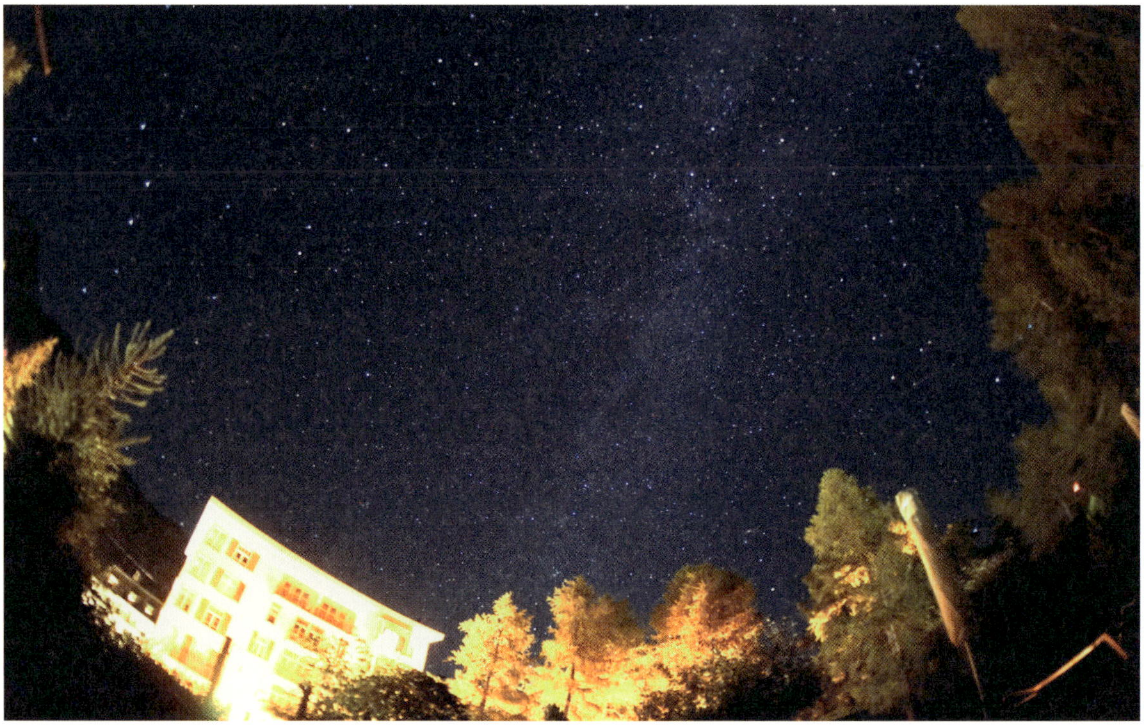

Figure 40. Milky Way photographed on August 22, 2020 at 10 pm at Waldhaus Sils Maria at about 1800 m above sea level.

very happy to receive it! Many good wishes to you and your family for a happy New Year in every way! With best regards. Yours, Gerhard Richter."

To conclude the section, I would like to return to Richter's fascination with large numbers. In the discussion of "25 colors" (Figure 3), it was mentioned that putting all possible color permutations of this image behind each other would give ten thousand times the diameter of our galaxy, the Milky Way shown in Figure 40.

Unlike Lego bricks, atoms of a given element and isotope are completely indistinguishable. Therefore, there is only one way to arrange 25 atoms of the same kind in a 5 × 5 grid as shown in Figure 41.

Our galaxy contains an estimated 10^{12} (one trillion) suns, and the entire observable universe consists of about 10^{12} galaxies and the number of atoms in the universe is estimated to be about 10^{85}.

Humans are made up of about 10^{28} atoms, most of which are hydrogen. The oceans of the world contain about 10^{47} atoms. Even assuming that our ancestors are evenly dissolved in the world's oceans, this means that each of us contains about $10^{28} \times 10^{28}/10^{47} = 10^{9}$, or one billion, atoms that have

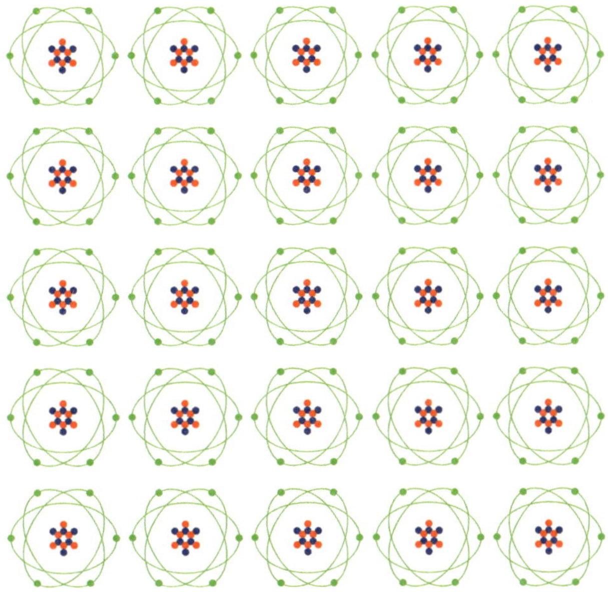

Figure 41. 25 atoms, here the carbon isotope ^{12}C. As atoms of the same chemical element and the same number of neutrons are indistinguishable, there is only one way to arrange those 25 atoms.

circulated before in Leonardo da Vinci, Mozart, Caesar, and others. This surprising realization is mitigated by the indistinguishability of the same isotopes, which, after all, are about as old as the universe at 13.8 billion years.

In any case, this comparison shows that matter alone does not account for everything.

D. Drive

One thing that has always fascinated me about Richter is his energy and drive. After all, I had offered to help him build the Münster pendulum. However, the University of Münster also has a physics institute with highly competent staff and a very good workshop, and the close proximity to the Dominican Church facilitates the project immensely. I would have loved to build the pendulum with and for Richter, but I was not unhappy that my colleagues in Münster had taken the matter into their hands.

All the more I was pleased about an urgent email message from Gerhard Richter of 07.12.2017:

From: Gerhard Richter <gerhard.richter@xxx.xxx> 07.12.2017 12:40

To: franz.giessibl@ur.de

Dear Mr. Gießibl, I do not know where and if you receive this letter, but I have to try in the matter of Münster. The press conference a week ago went well and the technical details were also discussed. Everything should go ahead as planned.

I can even understand that the physicists of the University of Münster are gripped by the ambition to master the task independently. And so the offer of your sphere as a gift was gratefully heard, but rather in such a way that it became clear that our sphere was not needed at all.

But this has suddenly changed, because the commissioner called me yesterday and reported that the physicists' group is now very interested in our sphere after all, in order to examine it. And therefore they are asking me to send the sphere to Münster, on loan, so to speak.

Since I am of course very interested in the success of the project, I would like to comply with this request.

And therefore I would like to ask you if you can agree to this, and if it would be possible to send the sphere directly to Münster - of course freight collect, to the debit of my FedEx account: xxxx-xxxx-x.

I am very sorry to bother you with this, but what can I do. And now I am very eager to hear your reply.

With good wishes and best regards

Yours Gerhard Richter

Our colleagues in Münster needed our sphere to test their pendulum drive, because the 48 kg pendulum sphere that had been contracted out to an external company was still being manufactured. In the end, they constructed a highly reliable drive and it was a great pleasure for us that we could contribute to the success and successful opening of the pendulum in June 2018 by lending our sphere.

E. Japan

Japan is a very strong player in my field of atomic-resolution microscopy, not least because the attention to detail is deeply rooted in Japanese culture. That's why I travel to Japan frequently and told Gerhard Richter now and then about these trips and my ascent of Mount Fuji in August 2014.

He then sent me a book about the artwork "14 Panes of Glass for Toyoshima, dedicated to futility," a hollow wooden prism in a bamboo forest on the small island of Toyoshima that contains 14 panes of glass dedicated to futility. Richter was even present at the opening of the artwork in Japan in 2015.

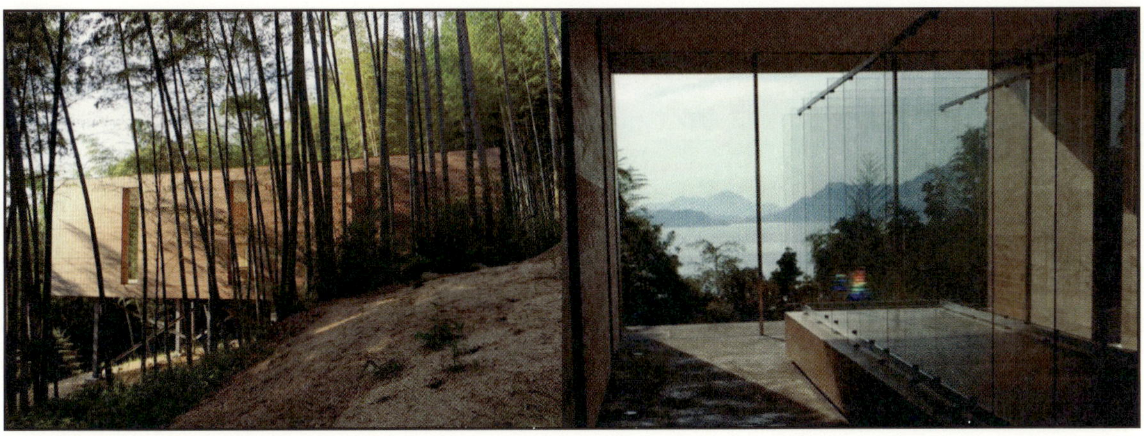

Figure 42. 14 Panes of Glass for Toyoshima, dedicated to futility. Gerhard Richter (2016).

F. Gray

Leonardo da Vinci worked on the Mona Lisa for 14 years. The work began in 1503 in Florence, he took it to Milan, Rome and France and finished it in 1517 (p. 475 in [2]). The painting is a condensate of the knowledge accumulated over his life in art and science. To study the interplay between facial expressions and emotions for his grotesque portraits, he walked the streets with a sketchpad on his belt, inviting passersby who served as models to lunch, telling them the craziest stories to make them laugh wildly. Their facial expressions burned into his memory and he drew them afterwards (p. 121 in [2]).

Probably every serious artist and scientist has an idea, a hunch, that accompanies him all his life and that he may not easily put into words nor explain from basic principles.

Some viewers who greatly appreciate Richter's photorealistic paintings have difficulties with the gray monochrome paintings and mirrors. The color gray must have gripped Richter for a very long and lasting time. In Corinna Belz' film *"Gerhard Richter, Painting"*, Richter amusingly recounts at 1 h 30 min the candidness of the American exhibition public. A group of young artists who had visited his first exhibition at Onnasch in New York in 1973 with photorealistic paintings and gray monochrome paintings noted: "Great, great show, but the gray ones, you know what it is? Bullshit!"

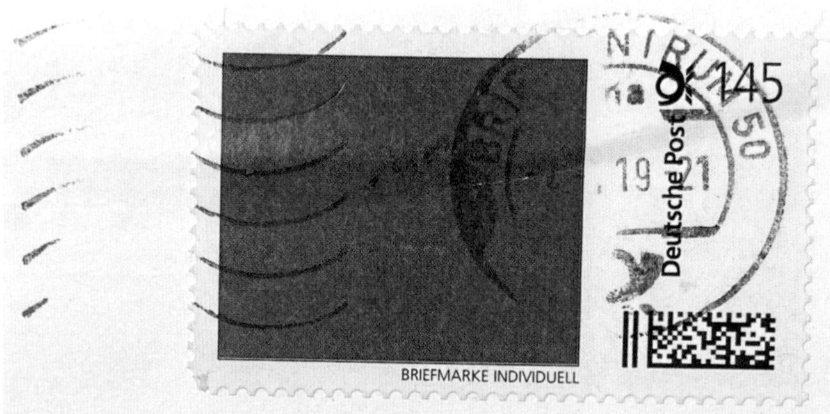

Figure 43. Gray stamp on a letter Gerhard Richter sent to the author.

Is it permissible that a painter, who paints a candle in such a way that you think you'll burn your fingers if you get too close, who paints an elegant woman descending the stairs in such a way that you can smell the perfume, just paints a gray monochrome painting? Is gray underestimated? In any case, the many visitors who have already visited the Münster pendulum with the two gray double mirrors show that the entire arrangement creates a strong effect.

Deutsche Post offers the possibility to design individualized stamps according to your own photos with the imprint "BRIEFMARKE INDIVIDUELL". Richter has even used a gray stamp on several of his letters and postcards to the author. (Figure 43).

G. Religion

On one of my visits to Cologne, Richter asked me if I, as a scientist, was religious. I told him about growing up in an Upper Bavarian village in the sixties, where almost all classmates were Catholic and the rest were Protestant, and church and village priest were a respected authority. Shortly before the turn of the millennium—around the birth of my sons—I had left the church, however, because I was very disappointed with the way the church was dealing with the abuse scandals that were becoming known at the time. Moreover, I have always been latently disturbed by what I perceive to be the church's limited acception of scientific progress, which also led to the fact that three centuries had to pass before Galileo Galilei was rehabilitated by the church.

The value of many things often only becomes apparent when they have disappeared. Hermann Lübbe identified religion as a reasonable contingency coping practice [49], that is, an aid for dealing with circumstances over which we are powerless. A private stroke of fate made me reconsider the decision to leave. In addition, our older son Severin attended the Regensburger Domspatzen high school, and we were often present when he sang with the Domspatzen choir at Sunday Mass in Regensburg Cathedral. Especially the Christmas masses were a lasting spiritual experience.

In the Pentling church near Regensburg, the emeritus dogmatics professor Wolfgang Beinert—long-time companion of Josef Ratzinger (Pope emeritus Benedict XVI) and very open to the natural sciences—held Sunday Mass and translated the Christian teachings in a captivating way into the present time, publishing weekly "Sunday meditations" even today via YouTube. About a decade after leaving, I applied for readmission to the Catholic Church—against the trend and even though the Church is not perfect. Self-recognition in line with the Bible verse from the Gospel according to John (8:7) "He who is without sin among you, let him be the first to cast a stone ..." contributed to this decision.

Jordan Peterson describes on page 186–187 of his book *"12 Rules For Life—An Antidote To Chaos"* [50] that Christianity introduced revolutionary ways of thinking about human liberation, such as the king and the beggar being on the same metaphysical level and equal before God and law. Under Roman law, it was still okay to throw live slaves to hungry lions for popular amusement. Christianity had made it clear very early on that women and men were equal, although the fleshing out of this equality is still a work in progress and the Protestant Church has already achieved more there. I find the symmetry between weak and powerful, poor and rich, sinful or pious captivating.

Wherever man is at work, mistakes happen. This is the case in the church, where from the Middle Ages to the present day, apart from all the good things done also the most serious injustices have been committed. In the natural sciences and even in the exact physical sciences, errors and mistakes were made in the past and in the presence. In physics, reproducibility is crucial: It does not count how renowned the propagator of a new theory or an experimental discovery is, what counts is whether it can be reproduced. In aviation, terrible accidents used to happen relatively frequently; fortunately, they are very rare today. Accidents or near-accidents in aviation are meticulously processed and published in public reports, even if no human is harmed and no property is damaged. Subsequently, multiple safety procedures are established to prevent the same accident from happening in the future. Concentration of power on individuals has often been found to be the cause of undesirable developments. This realization led to a certain reduction of the hierarchy from the flight captain to the co-pilot within the framework of cockpit resource management. If, for example, the co-pilot is of the opinion during a landing approach that a go-around procedure is indicated for safety reasons, then this is carried out, even if the captain does not share the assessment.

I would like to see the highly successful error-correction mechanisms of the natural sciences or aviation applied in the church as well; then the latter could communicate the beneficial teachings of Christ much more powerfully.

Richter was baptized Protestant as a child, but left the church as a young man (see p. 538 in [27]). In a 2004 interview with Jan Thorn-Prikker (p. 484

in [27]), Richter states: "I cannot believe in God, but I think the Catholic Church is great." To Jan Thorn-Prikker's question "What impresses you so much?" he replies: "It has shaped us, the Christian culture that exists in the world. I am grateful for the great art, the music, the most beautiful architecture, for the literature and philosophy, which is still thoroughly Catholic even when it preaches extreme atheism." In response to Jan Thorn-Prikker's inquiry, "Everything you revere in art comes from that thinking. There you gratefully accept religion, is that so?" Richter replies: "Well, it's more than that. When we had our two children baptized here in the [Cologne] cathedral, my attitude toward the church had changed a lot, I gradually realized what the church can offer, how much meaning it gives, how much support, comfort and security."

Thus, it is not surprising that Gerhard Richter created two important works for churches pro bono: the south window of the Cologne Cathedral and three altar windows of the monastery church of Tholey close to Saarbrücken. He also donated the two gray double mirrors for a pendulum in the profaned Dominican Church to the city of Münster.

The cathedral window is shown in Figure 44. It consists of about 11 500 square color panels in 72 ancient color schemes, which were distributed essentially at random. The number of possible combinations exceeds the number of atoms in the universe by a huge factor, even when partial mirror symmetry is taken into account. Numbering the six partial windows along the horizontal from 1 to 6, we see that 1 and 3, 2 and 5, and 4 and 6 are mirrored about the vertical axis, which makes the window appear much calmer than a completely random arrangement.

Gerhard Richter's cathedral window contains two seemingly opposite aspects of creation, which are found in the foundations of physics: chance and symmetry. Even if one firmly believes in a providence, one will not be able to deny that many events seem to be random. Quantum theory includes chance even with the uncertainty relations for position and momentum and for time and energy at its core. Symmetries are of paramount importance in mathematics and physics. One can also view the subtle arrangement of the cathedral windows as an expression of the symmetry between king and beggar. Therefore, the window fits very well in a cathedral.

Aesthetics is another possible reason for the symmetry of the cathedral windows. Mario Livio describes in his book *"The equation that couldn't be solved: How Mathematical Genius Discovered the Language of Symmetry"* [51] important aspects of symmetry in mathematics and science, but also the importance of mirror symmetry for human perception.

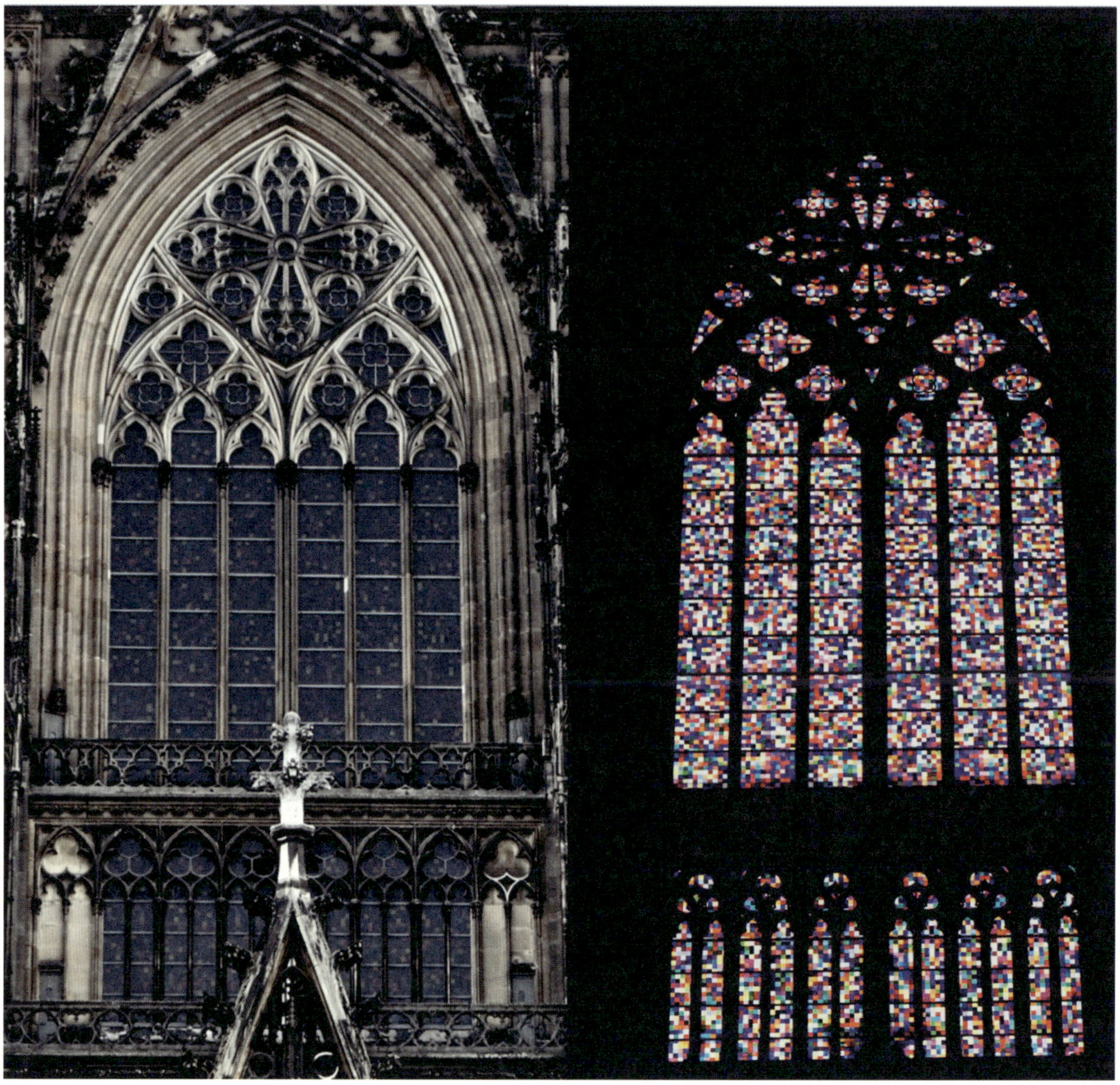

Figure 44. Cologne Cathedral window from the outside (left) and from the inside (right), designed by Gerhard Richter (2007). About 11 500 glass squares with a side length of 9,7 cm in 72 colors are arranged in six vertical lanes, with the distribution of colors between lanes 1-3, 2-5 and 4-6 mirrored around the vertical axis.

For this purpose, one can perform a simple experiment on the optical impression of mirrored structures. Gerhard Richter's work "4900 Colors, Version II (2007)" forms the basis for this. The original consists of 49 panels, each with 10 by 10 color fields and a size of 97 cm square. From these 49 plates in figure 45, 6 plates were selected at random and the left half of each plate with 5 by 10 color fields was lined up.

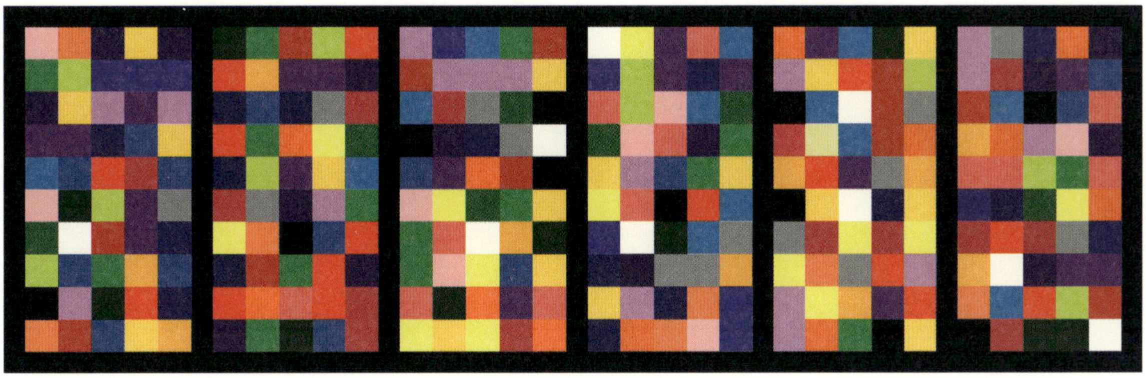

Figure 45. Excerpts from Gerhard Richter's work "4900 Farben, Version II (2007)", cropped and randomly arranged.

The random arrangement of six different color panels in Figure 45 seems somewhat restless. The mirroring of the left three panels around the vertical axis provides a calmer, but perhaps also somewhat boring image, shown in Figure 46.

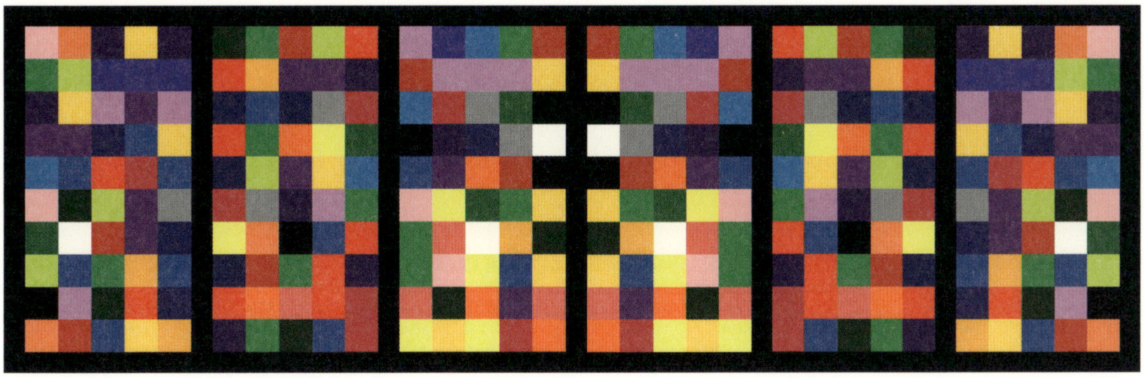

Figure 46. The left half of Figure 45, consisting of three panels, mirrored around the vertical axis to a symmetry 1 and 6, 2 and 5 followed by 3 and 4.

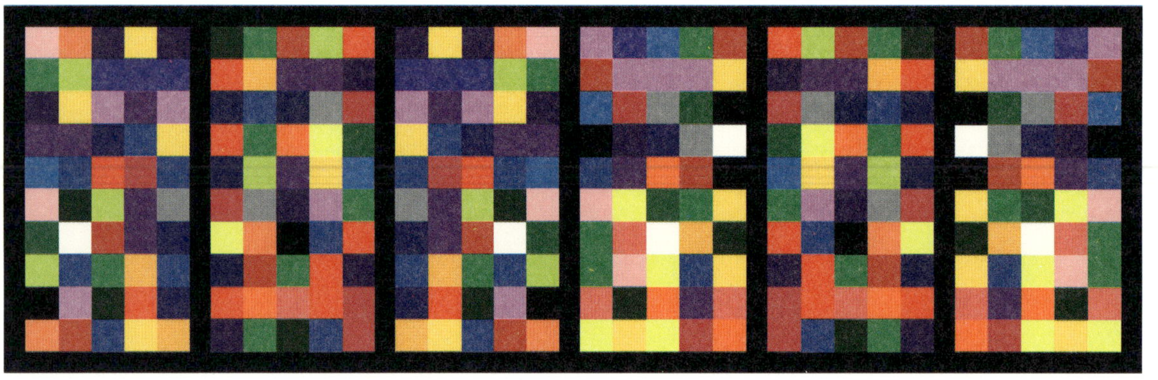

Figure 47. The left half of Figure 45, consisting of three panels, mirrored to a symmetry 1 and 3, 2 and 5 followed by 4 and 6.

Finally, Figure 47 shows the symmetry realized in the Cologne Cathedral window. Here, the mirror symmetry between panels 1 and 3, 2 and 5 as well as 4 and 6 presents itself in a subtle and interesting way.

Albert Einstein's quote "God does not play dice" turned against the statistical nature of quantum mechanics. Einstein did not believe the fuzziness and indeterminacy of the atomic world to be true and suggested that hidden parameters might yet grant determinacy to the world. Until today these parameters have not been found in modern physics and the Heisenberg uncertainty principle remains valid. Symmetries are however numerous in physics and thus the Cologne cathedral window shows with the coincidence and the symmetry two central pillars of creation.

X. Perspective

In section IV B it was already mentioned how the physicist Werner Heisenberg notes the limited expressive possibilities of language concerning the nature of quantum mechanics. The American psychotherapist and writer Irvin D. Yalom also concludes his book *"Love and its Executioner"* with a remark about the inadequacies of language with respect to the rendering of psychological realities. He quotes from Gustave Flaubert's *"Madame Bovary"*: "[T]he truth is that fullness of soul can sometimes overflow in utter vapidity of language, for none of us can ever express the exact measure of his needs or his thoughts or his sorrows; and human speech is like a cracked kettle on which we tap crude rhythms for bears to dance to, while we long to make music that will melt the stars."

Gerhard Richter does not even attempt to outline his paintings in words by drumming crude rhythms on cracked boilers, but instead declares: "My paintings are smarter than I am" [52], inviting us to explore his works for ourselves. The thousands of artworks created by him span a comprehensive and inspiring space for interpretation. He has overcome the past in painting and revived abstraction ("Overcoming the past in painting, revival of abstraction...", p. 811 in [53]).

A reflection of 20 years of interaction with Richter can be found on the south wall of my office at the University of Regensburg (Fig. 48). On the left is a reproduction of the memento mori "Candle with Skull (1983)." The center is taken by "Graphite (2004)", which depicts the central chemical element of life—carbon—in graphite modification. On the right hangs a reproduction of "Reader (1994)". The ensemble is a pictorial version of Emily Dickinson's poem

Death is a Dialogue between The Spirit and the Dust.
"Dissolve" says Death.
The Spirit "Sir I have another Trust."

interpreted as a consolation of one's own transience through the creation of something worth reading or viewing.

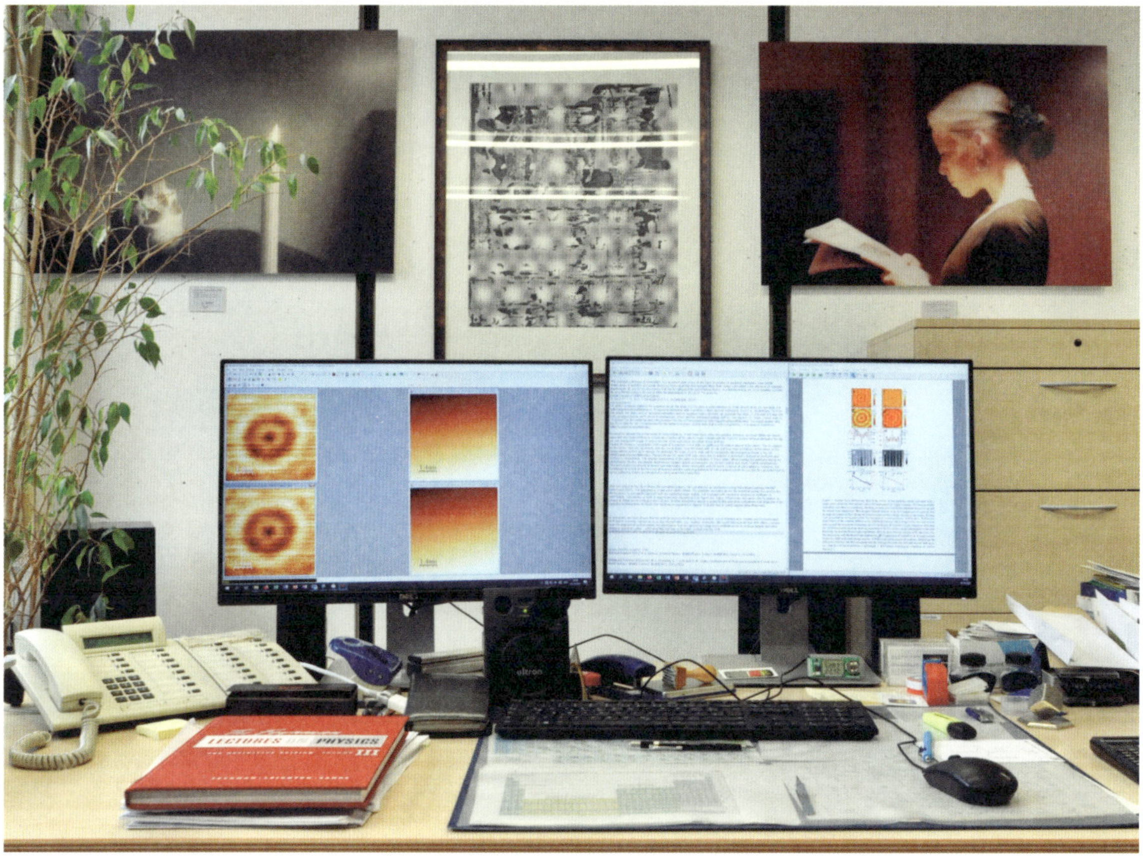

Figure 48. Office of the author at the University of Regensburg with "Graphite (2004)" (center) and reproductions of "Candle with Skull (1983)" (left) and "Reader (1994)" (right). The traces of an artificial atom can be seen on the computer screen as published in [54].

XI. Acknowledgments

I thank Gerhard Richter for two decades of exchange and inspiration, for the direct encounters, invitations, letters, cards, telephone calls, emails, for the works of art and books he gave me. I am very grateful to his office manager Mrs. Konstanze Ell for the always efficient and straightforward communication and so much help.

I also thank Mr. Walther König for his kind support to publish this book with its unusual genre between science and art.

Gerd Binnig opened the field of nanoscience and gave me the opportunity to participate from the beginning in this exciting exploration of new territory. He also allowed me to share in his extraordinary worldview and his confidence in tackling even daunting challenges with a courageous approach in the spirit of Nietzsche's Zarathustra. I further thank him for help and advice over many decades.

Jochen Mannhart enabled me to return from industry to academia. He gave me a trustworthy position at his chair at the University of Augsburg, granted me every freedom to realize the potential of my new invention, the force microscopy sensor, and supported the activities with much energy and enthusiasm.

A big thank you goes to the many students, PhD students, postdocs and visiting scientists with whom I was allowed to do research. Many of the projects were generously funded by the German Research Foundation.

The first ideas for this book arose during the preparation for a lecture on the occasion of the seminar Magic of Skiing in Aspen, Colorado, with Tom Crum in January 2017. Further impulses were provided by the preparation for the lecture in the Magnushaus at the German Physical Society in October 2019, initiated by Wolfgang Eberhardt. Further, I thank Ingo Meller and Ferdinand Evers for the opportunity to participate in an intercultural experiment at the Leipzig Academy of Visual Arts in November 2019, which also provided important inspiration for this book. Finally, I thank the Dietrich and Kienberger families of the Hotel Waldhaus in Sils Maria for the opportunity to speak about the topic at the Waldhaus in August 2020.

I sincerely thank the following people for many helpful suggestions to improve the manuscript: Gerd Binnig, Michael Hagner, Marina Libermann, Mario Livio, Peter Marinkovic, Ingo Meller, Kristian Hvidtfelt Nielsen,

Victoria Vesna. I am deeply indebted to these highly respected friends and colleagues for their many thoughtful suggestions, even though I did not implement all of them.

I would like to thank Anja Merkel and Florian Griesbeck, as well as the members of the machine shop of the Department of Physics for their help in constructing the Foucault pendulums, models and apparatus.

My wife Birgit greatly reduced her professional activities while our children were young, thus enabling me to experience the joys of a small family despite the long hours in research and teaching and the many trips. She also accompanied me to many of Richter's exhibition openings, let me travel alone to Cuba for 20 days to write the first draft of the manuscript, and made important suggestions for improving the manuscript. I would like to thank my sons Severin and Valentin for their vital communication of the life realities of those born around the turn of the millennium.

The first draft of the book was written in January 2021 at the Royalton Hicacos Hotel in Cuba. I am grateful to the hotel staff who provided a desk on a spacious balcony overlooking the palm garden and within earshot of the sound of the sea in a stunning atmosphere. Likewise, I thank the hotel guests, who were few in number due to the Corona pandemic but all the more interesting for it, for many conversations that provided further inspiration for this book project.

I thank Corinna Eindorfer and Anja Huber for copy editing and Anna Weindl for help with the word processing environment LaTeX.

I would like to thank the Regensburger Universitätsstiftung for their generous support of this book project.

List of illustrations

Literature

[1] G. Binnig, Aus dem Nichts – Über die Kreativität von Natur und Mensch, Piper München Zürich (1989).

[2] Walter Isaacson, Leonardo da Vinci—The Biography, Simon & Schuster, London, New York, Sydney, Toronto, New Delhi (2017).

[3] Armin Zweite, Das Denken ist beim Malen das Malen – Gerhard Richter – Leben und Werk. Schirmer-Mosel, Lenbachhaus (2019).

[4] F. J. Giessibl, S. Hembacher, H. Bielefeldt, J. Mannhart, Subatomic features on the silicon (111)-(7x7) surface observed by atomic force microscopy. *Science 289*, 422 (2000).

[5] H. J. Hug, M. A. Lantz, A. Abdurixit, P. J. A. van Schendel, R. Hoffmann, P. Kappenberger, A. Baratoff and reply by Franz J. Giessibl, Stefan Hembacher, Hartmut Bielefeldt, Jochen Mannhart, Subatomic Features in Atomic Force Microscopy Images, *Science 291*, 2509 (2001).

[6] Minghuang Huang, Martin Čuma, and Feng Liu, Seeing the Atomic Orbital: First-Principles Study of the Effect of Tip Termination on Atomic Force Microscopy, *Phys. Rev. Lett. 90* 256101 (2003).

[7] Dietmar Elger, Gerhard Richter, Maler. Dumont Verlag, Köln.

[8] G. Binnig, H. Rohrer, Nobel Lecture: Scanning tunneling microscopy—from birth to adolescence. *Rev. Mod. Phys. 59*, 615 (1987).

[9] Max Frisch, Homo faber. Suhrkamp, Frankfurt am Main (1957).

[10] Abbot et al., Observation of Gravitational Waves from a Binary Black Hole Merger. *Phys. Rev. Lett. 116*, 061102 (2016).

[11] G. Binnig, Atomic Force Microscope and Method for Imaging Surfaces with Atomic Resolution, US Patent No. 4,724,318 (1986).

[12] G. Binnig, C. F. Quate, Ch. Gerber, Atomic Force Microscope, *Phys. Rev. Lett. 56*, 930 (1986).

[13] F. J. Giessibl, Advances in Atomic Force Microscopy, *Reviews of Modern Physics 75*, 949 (2003).

[14] G. Binnig, H. Rohrer, Ch. Gerber, E. Weibel, 7x7 Reconstruction on Si(111) Resolved in Real Space. *Phys. Rev. Lett. 50*, 120 (1983).

[15] F.J. Giessibl, G. Binnig, Investigation of the (001) cleavage plane of potassium bromide with an atomic force microscope at 4.2 K in ultra-high vacuum. *Ultramicroscopy 42–44*, 281 (1992).

[16] G. Binnig, Force microscopy. *Ultramicroscopy 42–44*, 7 (1992).

[17] F. J. Giessibl, Atomic Resolution of the Silicon (111)-7x7 Surface by Atomic Force Microscopy. *Science 267*, 68 (1995).

[18] Rasterkraftmikroskop für "sauberes" Silizium, Frankfurter Allgemeine Zeitung, Natur und Wissenschaft, 22. Februar 1995, Seite N3.

[19] Dava Sobel, Longitude. Harper Collins (2000).

[20] Ferdinand Evers, Wissenschaft in der Lieferkette, *Frankfurter Allgemeine Zeitung*, 3. Februar 2021, Seite N4.

[21] Richard P. Feynman, Robert B. Leighton, and Matthew L. Sands. The Feynman Lectures on Physics. Reading, Mass: Addison-Wesley Pub. Co (1963).

[22] Dalai Lama, Die Welt in einem einzigen Atom, Theseus Verlag Berlin (2005).

[23] F. J. Giessibl, AFM's path to atomic resolution. *Materials Today 8*, 32 (2005).

[24] A. Extance, The atomic-force revolution, *Nature 555*, 545 (2018).

[25] Werner Heisenberg, Der Teil und das Ganze, Gespräche im Umkreis der Atomphysik. R. Piper & Co Verlag, München (1969).

[26] Hubertus Butin, Gerhard Richters "Erster Blick" im Kontext nanotechnologischer Visualisierungen. In Annerose Keßler, Isabelle Roth (Hg.) Objektivität und Imagination – Naturgeschichte in der Kunst des 20. und 21. Jahrhunderts, Transcript Verlag (2018).

[27] Dietmar Elger, Hans Ulrich Obrist (Hrsg.) Gerhard Richter, Text 1961 bis 2007. Schriften, Interviews, Briefe. Verlag der Buchhandlung Walther König, Köln (2008).

[28] Antonia Fehrenbach, Der Lotuseffekt, Axel Dielmann Verlag, Frankfurt (2008).

[29] Dietmar Elger, Sumi Hayashi, *Gerhard Richter—14 Panes of Glass for Toyoshima, dedicated to futility*. Wako Works of Art, Tokyo (2016).

[30] Gerhard Richter, *Patterns*. Thames and Hudson, London (2012).

[31] S. Hembacher, F.J. Giessibl, J. Mannhart, C.F. Quate, Revealing the hidden atom in graphite by low-temperature atomic force microscopy. *PNAS 100*, 12539 (2003).

[32] Zaoli Zhang, Wilfried Sigle, Fritz Phillipp, Manfred Rühle, Direct Atom-Resolved Imaging of Oxides and Their Grain Boundaries *Science 302*, 846 (2003).

[33] Tamarra Kemsley, May 23, 2013 https://www.natureworldnews.com/articles/2062/20130523/nobel-prize-winning-physicist-henrich-rohrer-dies-79.htm,

[34] Kristian Hvidtfelt Nielsen, Nanotech, Blur and Tragedy in Recent Artworks by Gerhard Richter. *Leonardo 41*, 484–492 (2008).

[35] Karl Popper, Alles Leben ist Problemlösen, Piper Verlag München (1994).

[36] W. Isaacson, Steve Jobs. Simon&Schuster, New York (2011).

[37] S. Hembacher, F. J. Giessibl, J. Mannhart, Force microscopy with light-atom probes, *Science 305*, 380 (2004).

[38] Th. Hofmann, X. Ren, A. J. Weymouth, D. Meuer, A. Liebig, A. Donarini, F. J. Giessibl, Evidence for temporary and local transition of sp^2 graphite-type to sp^3 diamond-type bonding induced by the tip of an atomic force microscope, https://arxiv.org/abs/2103.12660 (2021).

[39] J. Welker, F.J. Giessibl, Revealing the Angular Symmetry of Chemical Bonds by Atomic Force Microscopy. *Science 336*, 442 (2012).

[40] M. Emmrich, F. Huber, F. Pielmeier, J. Welker, T. Hofmann, M. Schneiderbauer, D. Meuer, S. Polesya, S. Mankovsky, D. Ködderitzsch, H. Ebert, F. J. Giessibl, Subatomic resolution force microscopy reveals internal structure and adsorption sites of small iron clusters, *Science 348*, 308 (2015).

[41] L. Gross, F. Mohn, N. Moll, P. Liljeroth, G. Meyer, The chemical structure of a molecule resolved by atomic force microscopy, *Science 325*, 1110 (2009).

[42] Mario Livio, Brilliant Blunders: From Darwin to Einstein–Colossal Mistakes by Great Scientists That Changed Our Understanding of Life and the Universe, Simon and Schuster, New York (2013).

[43] F. Huber, J. Berwanger, S. Polesya, S. Mankovsky, H. Ebert, F. J. Giessibl, Chemical bond formation showing a transition from physisorption to chemisorption. *Science 366*, 243 (2019).

[44] F. Charron, L' Experience du Pendule de Foucault, *L'Astronomie 45*, 457 (1931).

[45] Michael Hagner, Foucault's Pendel und wir. Anlässlich einer Installation von Gerhard Richter, Verlag der Buchhandlung Walther König, Köln (2021).

[46] Dieter Schwarz, Gerhard Richter. Zwei graue Doppelspiegel für ein Pendel in Münster, Verlag der Buchhandlung Walther König, Köln (2021).

[47] Max Frisch, Don Juan oder Die Liebe zur Geometrie. Komödie in fünf Akten. In: Stücke. Band 2. Suhrkamp, Frankfurt am Main (Erstausgabe der zweiten Fassung) (1962).

[48] Brief von Beethoven an Nikolaus Simrock in Bonn, Wien, 28. November 1820 (BGA 1418). Beethoven-Haus Bonn, Sammlung H. C. Bodmer, HCB Br 233 (1820).

[49] Hermann Lübbe, Religion nach der Aufklärung, Wilhelm Fink Verlag, München (2004).

[50] Jordan B. Peterson, 12 Rules For Life—An Antidote to Chaos, Random House, Canada (2018).

[51] Mario Livio, The equation that couldn't be solved: How Mathematical Genius Discovered the Language of Symmetry, Simon and Schuster, New York (2006).

[52] Viktoria von Flemming, „Gerhard Richter – Meine Bilder sind klüger als ich", Dokumentationsfilm Norddeutscher Rundfunk (1992).

[53] Peter Watson, The German Genius, Harper Collins, New York (2010).

[54] Fabian Stilp, Andreas Bereczuk, Julian Berwanger, Nadine Mundigl, Klaus Richter, Franz J. Giessibl, Very weak bonds to artificial atoms formed by quantum corrals. *Science 372*, 1196 (2021).